Soil Sampling, Preparation, and Analysis

BOOKS IN SOILS, PLANTS, AND THE ENVIRONMENT

Soil Biochemistry, Volume 1, edited by A. D. McLaren and G. H. Peterson
Soil Biochemistry, Volume 2, edited by A. D. McLaren and J. Skujiņš
Soil Biochemistry, Volume 3, edited by E. A. Paul and A. D. McLaren
Soil Biochemistry, Volume 4, edited by E. A. Paul and A. D. McLaren
Soil Biochemistry, Volume 5, edited by E. A. Paul and J. N. Ladd
Soil Biochemistry, Volume 6, edited by Jean-Marc Bollag and G. Stotzky
Soil Biochemistry, Volume 7, edited by G. Stotzky and Jean-Marc Bollag
Soil Biochemistry, Volume 8, edited by Jean-Marc Bollag and G. Stotzky

Organic Chemicals in the Soil Environment, Volumes 1 and 2, edited by C. A. I. Goring and J. W. Hamaker
Humic Substances in the Environment, M. Schnitzer and S. U. Khan
Microbial Life in the Soil: An Introduction, T. Hattori
Principles of Soil Chemistry, Kim H. Tan
Soil Analysis: Instrumental Techniques and Related Procedures, edited by Keith A. Smith
Soil Reclamation Processes: Microbiological Analyses and Applications, edited by Robert L. Tate III and Donald A. Klein
Symbiotic Nitrogen Fixation Technology, edited by Gerald H. Elkan
Soil-Water Interactions: Mechanisms and Applications, Shingo Iwata and Toshio Tabuchi with Benno P. Warkentin
Soil Analysis: Modern Instrumental Techniques, Second Edition, edited by Keith A. Smith
Soil Analysis: Physical Methods, edited by Keith A. Smith and Chris E. Mullins
Growth and Mineral Nutrition of Field Crops, N. K. Fageria, V. C. Baligar, and Charles Allan Jones
Semiarid Lands and Deserts: Soil Resource and Reclamation, edited by J. Skujiņš
Plant Roots: The Hidden Half, edited by Yoav Waisel, Amram Eshel, and Uzi Kafkafi
Plant Biochemical Regulators, edited by Harold W. Gausman
Maximizing Crop Yields, N. K. Fageria
Transgenic Plants: Fundamentals and Applications, edited by Andrew Hiatt
Soil Microbial Ecology: Applications in Agricultural and Environmental Management, edited by F. Blaine Metting, Jr.
Principles of Soil Chemistry: Second Edition, Kim H. Tan

Water Flow in Soils, edited by Tsuyoshi Miyazaki
Handbook of Plant and Crop Stress, edited by Mohammad Pessarakli
Genetic Improvement of Field Crops, edited by Gustavo A. Slafer
Agricultural Field Experiments: Design and Analysis, Roger G. Petersen
Environmental Soil Science, Kim H. Tan
Mechanisms of Plant Growth and Improved Productivity: Modern Approaches, edited by Amarjit S. Basra
Selenium in the Environment, edited by W. T. Frankenberger, Jr., and Sally Benson
Plant–Environment Interactions, edited by Robert E. Wilkinson
Handbook of Plant and Crop Physiology, edited by Mohammad Pessarakli
Handbook of Phytoalexin Metabolism and Action, edited by M. Daniel and R. P. Purkayastha
Soil–Water Interactions: Mechanisms and Applications, Second Edition, Revised and Expanded, Shingo Iwata, Toshio Tabuchi, and Benno P. Warkentin
Stored-Grain Ecosystems, edited by Digvir S. Jayas, Noel D. G. White, and William E. Muir
Agrochemicals from Natural Products, edited by C. R. A. Godfrey
Seed Development and Germination, edited by Jaime Kigel and Gad Galili
Nitrogen Fertilization in the Environment, edited by Peter Edward Bacon
Phytohormones in Soils: Microbial Production and Function, W. T. Frankenberger, Jr., and Muhammad Arshad
Handbook of Weed Management Systems, edited by Albert E. Smith
Soil Sampling, Preparation, and Analysis, Kim H. Tan

Additional Volumes in Preparation

Soil Erosion, Conservation, and Rehabilitation, edited by Menachem Agassi

Photoassimilate Distribution of Plants and Crops: Source–Sink Relationships, edited by Eli Zamski and Arthur A. Schaffer

Plant Roots: The Hidden Half, Second Edition, Revised and Expanded, edited by Yoav Waisel, Amram Eshel, and Uzi Kafkafi

Soil Sampling, Preparation, and Analysis

Kim H. Tan
The University of Georgia
Athens, Georgia

Marcel Dekker, Inc.　　　New York•Basel•Hong Kong

Library of Congress Cataloging-in-Publication Data

Tan, Kim H. (Kim Howard)
  Soil sampling, preparation, and analysis / Kim H. Tan.
    p. cm. — (Books in soils, plants, and the environment ; 45)
  Includes bibliographical references (p.   ) and index.
  ISBN 0-8247-9675-6 (hardcover : alk. paper)
  1. Soils—Analysis. 2. Soils—Sampling. I. Title. II. Series.
S593.T35 1995
631.4'1'0287—dc20   95-31146
                                                              CIP

The publisher offers discounts on this book when ordered in bulk quantities. For more information, write to Special Sales/Professional Marketing at the address below.

This book is printed on acid-free paper.

Copyright © 1996 by MARCEL DEKKER, INC. All Rights Reserved.

Neither this book nor any part may be reproduced or transmitted in any form or by any means, electronic or mechanical, including photocopying, microfilming, and recording, or by any information storage and retrieval system, without permission in writing from the publisher.

MARCEL DEKKER, INC.
270 Madison Avenue, New York, New York 10016

Current printing (last digit):
10 9 8 7 6 5 4 3 2 1

PRINTED IN THE UNITED STATES OF AMERICA

PREFACE

Many books now available on soil chemical analysis are bulky and voluminous, and all kinds of extraneous analyses are discussed. Most of the books stress the importance of the analytical procedures. Less emphasis is placed on proper sampling, handling, and preparation of soils for analysis, and the reader is left to contemplate arbitrary sampling and preparation procedures. Not only will the overstress on analytical procedures increase the chances of creating serious errors, but the user also is frequently very confused about the method of analysis to select among the numerous alternative procedures discussed in those books. Therefore, this book is written with the purpose to present in one volume the methods of soil sampling, preparation, and analysis that are most commonly used in soils laboratories in the United States as well as in many laboratories abroad. Errors, variability, and accuracy of sampling and analysis are discussed in detail. The intent is to provide a new and modernized version of the classical book by M. L. Jackson, *Soil Chemical Analysis*, Prentice-Hall, Inc., 1960, which is now not only outdated but also out of print.

The new book presents the most frequently employed, and simplest, procedures for analyzing soils and plant material today, useful in research and instruction in soil morphology, classification and genesis, soil chemistry, soil fertility and plant nutrition, soil mineralogy, soil biochemistry, soil and water conservation, forest and range soils, soil ecology, and soil testing. Scientists, scholars, professionals, growers, and students in extension, horticulture, nurseries, and other related fields and sciences in need of soil and plant analysis may find a simple and easy reference on soil analysis in this handy one-volume treatise. The book is composed of 17 chapters, and can be divided into three parts, e.g., chapters 1 to 4 covering the principles of soil sampling, preparation, errors and variability of analysis, and general purpose of soil chemical analysis; chapters 5 through 12 describing methods of analysis of soil properties and constituents; and chapters 13 through 17 examining the proper use and handling of the instruments mentioned in the preceding section of methods of analysis, or the application and handling of the most essential instruments commonly used today in a soils laboratory.

Accordingly, chapter 1 discusses types, mechanics, and accuracy of sampling. The amount and number of samples desired for accuracy in analysis are highlighted. Sample preservation and preparation for analysis are presented in chapter 2. Chapter 3 examines errors and variability of soil analysis and rejection of results. Chapter 4 discusses the general purpose and use of soil analysis. A clear distinction is made between analysis for soil testing, and that for pedology, soil chemistry, soil physics, and soil fertility. Some overlap between soil testing and soil fertility is addressed.

Chapters 5 through 12 present specific methods for analysis of soils properties and constituents; each starts with the definition and principles of the soil property or soil constituent and its analysis. In this second section, analytical procedures for determination of selected physical or physico-chemical properties of soils are covered first, which are then followed by methods of analysis of chemical properties of soils. Therefore, determination of soil water, composed of measurements of total available water and soil moisture tension, is presented in chapter 5. The three major methods for particle size distribution analysis, or soil texture determination, are covered in chapter 6. Chapter 7 comprises procedures for measurement of soil density and total pore space. Two types of the most common methods for determination of bulk density are highlighted, clod method and bulk density measurement of disturbed soil. The latter has important applications in both greenhouses and nurseries, where potting soils are used. The two most common methods, the indicator or field method and potentiometric method, for soil pH measurement are presented in chapter 8. Interpretations with respect to water and buffered pH, lime requirement, and lime potential are included. Chapter 9 covers major methods for determination of CEC, ECEC, CEC_p, and CEC_t, including two common methods for determination of exchangeable H^+ ions. The preference for using the NH_4OAc method for CEC determination is discussed. Determination of the essential macronutrient elements for plant growth in chapter 10 includes procedures for analysis of NH_4-N, NO_3-N, and NO_2-N, and both total and available or extractable P, K, Ca, Mg, and S. Analysis of microelements, including Al and Si, is described in a similar fashion in chapter 11. Chapter 12 discusses methods of analysis of soil organic matter, which includes organic-C determination, and determination of soil and aquatic humic matter with their chemical properties.

The third section of the book, chapters 13 through 17, addresses the principles and uses of spectrophotometry; flame photometry, including atomic absorption and plasma emission spectroscopy; infrared spectroscopy; x-ray diffraction analysis; and differential thermal analysis. A wealth of IR spectrograms, XRD diffractograms, and DTA thermograms of clay minerals and humic acids is included for easy reference.

I want to thank Dr. Harry A. Mills, Professor of Horticulture, The University of Georgia, and Dr. J. Benton Jones, Jr., Director, Macro-Micro International, Inc., Athens, GA, for their review and criticism. Appreciation is extended to Ms. Nickie

Whitehead for editing the manuscript and to Dr. Juan Carlos Lobartini and Mr. John Rema for their assistance with the laser printer in the development of this book. Thanks are also extended to the various publishers, scientific societies, and fellow scientists who gave permission to reproduce their materials. Finally, I want to acknowledge the patience and understanding of my wife, Yelli, who always supported me with great enthusiasm and a lot of encouragement.

Kim H. Tan

CONTENTS

Preface ... iii

Chapter 1 PRINCIPLES OF SOIL SAMPLING ... 1

  1.1 General Principles ... 1
  1.2 Soil Concept for Sampling ... 2
  1.3 Types of Soil Sampling ... 5

      1.3.1 Simple Random Sampling ... 6
      1.3.2 Systematic Sampling ... 6
      1.3.3 Stratified Sampling ... 7
      1.3.4 Compositing ... 8

  1.4 Mechanics of Sampling ... 9

      1.4.1 Shallow Sampling ... 9
      1.4.2 Deep Sampling ... 10
      1.4.3 Soil Profile Sampling ... 10
      1.4.4 Time of Sampling ... 11

  1.5 Size and Accuracy of Sampling ... 12

Chapter 2 SAMPLE PREPARATION 17

2.1 Sample Preservation 17
2.2 Sample Storage 20
2.3 Sample Preparation 21
2.4 Grinding and Sieving 21

    2.4.1 Grinding 21
    2.4.2 Sieving 22
    2.4.3 Errors in Grinding and Sieving 23

2.5 Reduction in Sample Size 25

Chapter 3 ERRORS AND VARIABILITY OF SOIL ANALYSIS 27

3.1 Measure of Accuracy 27
3.2 Measure of Precision 28
3.3 Measure of Bias 30
3.4 Types of Errors 32

    3.4.1 Determinate Errors 32
    3.4.2 Indeterminate Errors 34

3.5 Rejection of Results 35

    3.5.1 Checking for Accuracy 35
    3.5.2 Checking for Precision 38
    3.5.3 Checking for Bias 39

3.6 Statement of Results 40

Contents

Chapter 4 METHODS OF SOIL CHEMICAL ANALYSIS	42
4.1 Purpose and Use	42
4.2 Soil Testing	43
4.2.1 Monitoring	44
4.2.2 Determination of Nutrient Status	45
4.2.3 Plant Testing	45
4.2.4 Tissue Testing	47
4.3 Soil Analysis in Pedology and Soil Chemistry	48
4.3.1 Traditional Soil Analysis	48
4.3.2 Isolation and Analysis of Soil Constituents	49
4.3.3 Micropedological Analysis	49
4.3.4 Analysis of Soils *in Situ*	50
4.4 Types of Soil Chemical Analysis	51
4.4.1 Qualitative and Quantitative Analysis	51
Chapter 5 DETERMINATION OF SOIL WATER	56
5.1 Definitions and Principles	56
5.2 Determination of Total Soil Water Content	59
5.2.1 Direct Determination of Water Content	59
5.2.2 Indirect Determination of Water Content	61

5.3 Determination of Available Water Content   62

    5.3.1 Pressure-Plate Method   62
    5.3.2 Alternative Methods   65

5.4 Determination of Soil Moisture Tension   68

Chapter 6 DETERMINATION OF SOIL TEXTURE   73

6.1 Definition and Principles   73
6.2 Hydrometer Method   78

    6.2.1 Dispersion Reagent   79
    6.2.2 Procedure   79

6.3 Centrifuge Method   81

    6.3.1 Reagents   81
    6.3.2 Procedure   82

6.4 Pipette Method   84

    6.4.1 Procedure   84

Chapter 7 DETERMINATION OF SOIL
                DENSITY AND POROSITY   86

7.1 Definition and Principles   86
7.2 Measurement of Bulk Density   88

    7.2.1 Bulk Density Measurement of
           Disturbed Soil   88

Contents xi

    7.2.2 Clod Method   90

  7.3 Measurement of Particle Density   91

    7.3.1 Method Using a Graduated Cylinder   91
    7.3.2 Method Using a Volumetric Flask   93

  7.4 Determination of Pore Spaces   94

Chapter 8 SOIL pH MEASUREMENT   96

  8.1 Definition and Principles   96
  8.2 Indicator or Colorimetric Method   99
  8.3 Potentiometric Method   103
  8.4 Factors Affecting pH Measurements   106

    8.4.1 Suspension Effect   106
    8.4.2 Dilution Effect   106
    8.4.3 Sodium Effect   107

  8.5 Water pH Versus Buffer pH   107
  8.6 Determination of Lime Requirement   109
  8.7 Determination of Lime Potential   112

Chapter 9 CATION EXCHANGE CAPACITY DETERMINATION   114

  9.1 Definition and Principles   114
  9.2 Factors Affecting CEC Determination   117

    9.2.1 pH of the Leaching (Displacing) Solution   117
    9.2.2 Nature of the Exchange Complex   118

9.2.3 Nature and Concentration of the Leaching Solution	119
9.2.4 Presence of Undesirable Interactions	121
9.2.5 Analytical Approach	121
9.2.6 Limitations of the Analysis	122
9.3 Methods of CEC Determination	123
9.3.1 Ammonium Acetate (NH_4OAc) Method	123
9.3.2 Alternative Method	126
9.3.3 Determination of ECEC	127
9.3.4 Determination of CEC_p	128
9.3.5 Determination of CEC_t	130
9.4 Methods for Determination of Exchangeable H^+	132
9.4.1 Barium Chloride-TEA Method	132
9.4.2 Potassium Chloride (KCl) Method	133
Chapter 10 DETERMINATION OF MACROELEMENTS	135
10.1 Determination of Soil Nitrogen	135
10.1.1 Forms of Nitrogen in Soils	135
10.1.2 Determination of Total N	138
10.1.3 Kjeldahl Method	139
10.1.4 Determination of Exchangeable NH_4^+	148
10.1.5 Determination of NO_3^-	149
10.1.6 Determination of NO_2^-	152

Contents

10.2 Determination of Soil Phosphorus	153
10.2.1 Forms of Phosphorus in Soils	153
10.2.2 Determination of Total P	159
10.2.3 Determination of Organic P	162
10.2.4 Determination of Available P	165
10.3 Determination of Potassium	170
10.3.1 Forms of Potassium in Soils	170
10.3.2 Determination of Total K	173
10.3.3 Determination of Available K	174
10.4 Determination of Sodium	175
10.4.1 Forms of Sodium in Soils	175
10.4.2 Determination of Total Na	177
10.4.3 Determination of Exchangeable Na	178
10.5 Determination of Calcium	178
10.5.1 Forms of Calcium in Soils	178
10.5.2 Determination of Total Ca	180
10.5.3 Determination of Exchangeable Ca	180
10.5.4 Determination of Dilute Acid Soluble Ca	180
10.6 Determination of Magnesium	181
10.6.1 Forms of Magnesium in Soils	181
10.6.2 Determination of Total Mg	182
10.6.3 Determination of Exchangeable Mg	182
10.6.4 Determination of Dilute Acid Soluble Mg	182

10.7 Determination of Sulphur ... 183

    10.7.1 Forms of Sulphur in Soils ... 183
    10.7.2 Determination of Total S ... 186
    10.7.3 Determination of Inorganic S ... 187

Chapter 11 DETERMINATION OF MICROELEMENTS ... 189

11.1 Determination of Aluminum in Soils ... 189

    11.1.1 Forms of Aluminum in Soils ... 189
    11.1.2 Determination of Total Al ... 191
    11.1.3 Determination of Exchangeable Al ... 193

11.2 Determination of Iron in Soils ... 194

    11.2.1 Forms of Iron in Soils ... 194
    11.2.2 Determination of Total Fe ... 196
    11.2.3 Determination of Exchangeable Fe ... 196
    11.2.4 Determination of Available Fe ... 197
    11.2.5 Determination of Free and Amorphous Fe ... 198

11.3 Determination of Silicon in Soils ... 199

    11.3.1 Forms of Silicon in Soils ... 199
    11.3.2 Determination of Total Si ... 202

11.4 Determination of Manganese in Soils ... 204

    11.4.1 Forms of Manganese in Soils ... 204
    11.4.2 Determination of Total Mn ... 205
    11.4.3 Determination of Exchangeable Mn ... 206
    11.4.4 Determination of Available Mn ... 207

11.5　Determination of Copper in Soils	207
11.5.1　Forms of Copper in Soils	207
11.5.2　Determination of Total Cu	209
11.5.3　Determination of Exchangeable Cu	210
11.5.4　Determination of Available Cu	211
11.6　Determination of Boron in Soils	211
11.6.1　Forms of Boron in Soils	211
11.6.2　Determination of Total B	213
11.6.3　Determination of Available B	215
11.7　Determination of Molybdenum in Soils	216
11.7.1　Forms of Molybdenum in Soils	216
11.7.2　Determination of Total Mo	218
11.7.3　Determination of Available Mo	218
11.8　Determination of Zinc in Soils	219
11.8.1　Forms of Zinc in Soils	219
11.8.2　Determination of Total Zn	220
11.8.3　Determination of Exchangeable Zn	221
11.8.4　Determination of Available Zn	222
Chapter 12　DETERMINATION OF SOIL ORGANIC MATTER	223
12.1　Definition and Principles	223
12.2　Determination of Total Organic C	226
12.2.1　Dry Combustion Method	226
12.2.2　Alternative Method	228

12.3 Determination of Oxidizable Organic C	229
12.3.1 Walkley-Black Wet Combustion Method	231
12.4 Determination of Humic Matter	232
12.4.1 Extraction of Soil Humic Matter	234
12.4.2 Extraction of Aquatic Humic Matter	236
12.4.3 Determination of Total Acidity of Humic Matter	238
12.4.4 Determination of Carboxyl and Phenolic-OH Groups	239

Chapter 13 SPECTROPHOTOMETRY AND COLORIMETRY   242

13.1 Definition and Principles	242
13.1.1 Absorption Spectrophotometry	243
13.1.2 Principles of Light Absorption	244
13.1.3 Law of Lambert and Bouger	245
13.1.4 Beer's Law	246
13.1.5 Law of Lambert and Beer	246
13.2 Visible Light and UV Spectrophotometers	247
13.2.1 Single Beam and Double Beam Spectrophotometers	249
13.2.2 Source of Visible Radiation	250
13.2.3 Source of Ultraviolet Radiation	250
13.2.4 Monochromators	251
13.2.5 Detectors	254

Chapter 14 FLAME PHOTOMETRY — 255

14.1 Definition and Principles — 255
14.2 Atomic Absorption Spectroscopy — 256

    14.2.1 Principles of Atomic Absorption — 257
    14.2.2 Instrumentation — 258
    14.2.3 Samples and Sample Preparation — 265
    14.2.4 Interferences — 266

14.3 Atomic Emission Spectroscopy — 269

    14.3.1 Principles of Atomic Emission — 269
    14.3.2 Emission Spectrographic Methods — 271
    14.3.3 Plasma Emission Spectroscopy — 274

14.4 Detection Limits of Atomic Absorption and Plasma Emission Spectroscopy — 276

Chapter 15 INFRARED SPECTROSCOPY — 278

15.1 Definition and Principles — 278
15.2 Origin of Spectra — 280
15.3 Instrumentation and Sources of Infrared Radiation — 284

    15.3.1 Primary Source of Radiation — 284
    15.3.2 Detectors — 284

15.4 Sample Preparation — 285

    15.4.1 Gas Samples — 285
    15.4.2 Liquid Samples — 285
    15.4.3 Solid Samples — 285

15.5 Infrared Analysis of Humic Matter   286

    15.5.1 Procedures   286
    15.5.2 Infrared Spectra of Humic Matter   287
    15.5.3 Methylation of Humic Acids   291

15.6 Infrared Analysis of Soil Clays   292

    15.6.1 Materials and Pretreatments   292
    15.6.2 Procedures   292
    15.6.3 Infrared Spectra of Soil Clays   294

Chapter 16 X-RAY DIFFRACTION ANALYSIS   299

16.1 Definition and Principles   299
16.2 Instrumentation and Principles of
      X-ray Diffraction   303

    16.2.1 Sources of X-ray Radiation   303
    16.2.2 Principles of X-ray Diffraction   305

16.3 Qualitative X-ray Diffraction
      Spectrometry   308

    16.3.1 Pretreatments   308
    16.3.2 Mounting Techniques of Samples
           in XRD   311
    16.3.3 Preparation of Oriented Glass
           Slide Samples   312
    16.3.4 Preparation of Random Powder
           Samples   313
    16.3.5 Diffractograms of Clay Minerals   313

Contents

16.4 Quantitative X-ray Diffraction Spectrometry   327

    16.4.1 Quantitative Interpretations of Diffractograms   328

Chapter 17 DIFFERENTIAL THERMAL ANALYSIS   329

17.1 Definition and Principles   329
17.2 Instrumentation   332
17.3 DTA of Soil Minerals and Organic Matter   335

    17.3.1 Sample Preparation   335
    17.3.2 Sample Size   345
    17.3.3 Sample Packing   347
    17.3.4 Furnace Atmosphere   348
    17.3.5 Heating Rate   349

17.4 Qualitative Interpretation of DTA of Soil Minerals   350
17.5 Qualitative Interpretation of DTA of Organic Matter   357
17.6 Quantitative Interpretation of DTA of Soil Minerals   358

    17.6.1 Preparation of a Standard Curve   360

Appendix A Fundamental Constants   363
Appendix B Greek Alphabet   364
Appendix C Atomic Weights of the Major Elements in Soils   365
Appendix D X-ray Diffraction 2θ d-Spacing Conversion Table   368

Appendix E U.S. Weights and Measures	369
Appendix F International System of Units (SI);	371
Factors for Converting U.S. Units	
into SI Units	371
References and Additional Readings	375
Index	392

CHAPTER 1

PRINCIPLES OF SOIL SAMPLING

1.1 GENERAL PRINCIPLES

Before any soil chemical analysis can be performed, it is necessary to procure a test sample that will represent the soil under investigation and to prepare the test sample for analysis. Obtaining a truly representative sample of the soil under observation is not a simple task. Frequently sampling errors are commonly much greater than analytical errors (Reed and Rigney, 1947). The extent to which the result of an analysis identifies a real characteristic of the whole soil depends upon the accuracy of sampling, and efforts spent in precision analysis are wasted when the sample analyzed is not representative at all.

When the sample analyzed is not representative, the result of the soil chemical analysis will yield a value that does not necessarily describe the property of the soil as a whole. Rather, it delineates specifically a characteristic in the small test sample analyzed. According to Cline (1945), this analytical value can serve as an accurate description of the soil property under investigation only if:

1. The gross sample accurately represents the whole soil from which it was taken,

2. No changes occur in the gross and subsamples prior to analysis,
3. The subsample analyzed represents the gross sample accurately.
4. The analysis determines a true value of the soil characteristic under investigation.

Of the four points, the first is the most troublesome because of soil variability. Differences in soil composition and properties are present not only from region to region but also from one part of a given field to another, and frequently within very short distances of sampling sites. The second and third requirements can be controlled fairly well during sample preparation. The fourth is primarily related to the selection of the proper analytical procedure. Frequently, selecting the right procedure is more difficult than performing the actual analysis.

1.2 SOIL CONCEPT FOR SAMPLING

Soil is defined in soil science as an independent body in nature with a unique morphology from the surface down to the parent material as expressed by the soil profile. Soil is the product of biochemical weathering of the parent material, and its formation is influenced by the soil formation factors: climate, organisms, parent material, relief, and time (Jenny, 1941; Brady, 1990).

An individual soil body, briefly called soil, occurs in the landscape side by side with other soils like the pieces in a jigsaw puzzle. In many cases, an individual soil occupies a very large area or volume. The smallest representative unit of such a body is called a pedon. A pedon has three dimensions and is comparable in some way to a unit cell of a crystal. One soil body consists of contiguous similar pedons, collectively called polypedons. A pedon may range in area from 1 to 2 m^2, and it is bordered on all four sides by vertical sections, or soil profiles

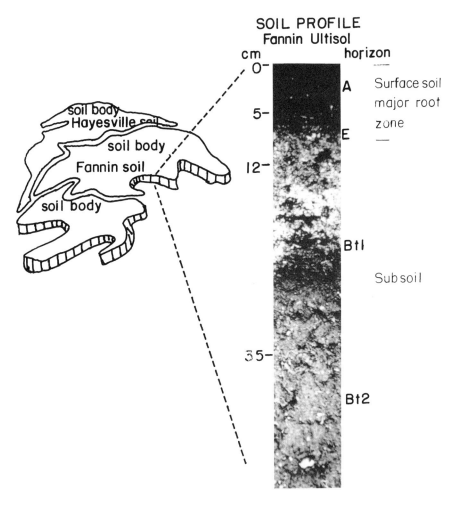

Figure 1.1 The relationship of a soil profile to a soil body.

(Figure 1.1). A soil profile, which extends from the surface down to the parent material, is composed of several horizontal layers called soil horizons. Six main groups of horizons, known as master horizons, are recognized. They are designated by the symbols O, A, E, B, C, and R (Soil Survey Staff, 1975). Each of the horizons may differ with respect to soil texture, structure, permeability, biological activity, and other attributes of importance in soil formation, fertility and crop production.

It is almost an impossible task to bring the whole soil body in the laboratory, and the most convenient way for chemical analysis is to take samples from soil pedons. The pedon or part of the pedon can be referred to as a *sampling volume*. Plant growth and performance are affected by the entire soil volume and not by soil area. Each soil volume from which a sample is drawn represents a population of soil constituents, e.g., soil moisture, organic matter, nutrients, and primary and secondary minerals. The latter, known as *individuals* in statistics, may differ in concentration and composition both vertically and horizontally in the pedon. Each soil sample is composed of a group of closely associated soil constituents, such as a soil core or soil slice of given dimension, and for convenience can be regarded as a *sampling unit*. Sampling units may also vary among themselves both horizontally and vertically. For the purpose of sampling, each sampling unit may be treated as an individual (Cline, 1944, 1945).

The gross sample, consisting of one to more sampling units, is considered a single soil population. Its validity as a representative of the soil as a whole depends upon the following criteria:

1. The degree of homogeneity of the sampling volume. The less homogeneous the sampling volume is, the greater will be the sampling error.
2. The method of sampling. Samples can be selected from a pit with or without exposing the soil profile or they may be

Principles of Soil Sampling

drawn from road banks. They can be taken with a spade or an auger. Samples obtained from soil profiles may be more representative than auger samples, as will be discussed in more detail later.

3. The number of sampling units contributing to the gross sample. The use of several sampling units instead of only one sampling unit may yield a gross sample providing a closer approximation of the soil. Several small sampling units, each drawn from a homogeneous part of the pedon, yield a better representation of the whole soil than does a single large sample.

In sampling pedons, the rule of homogeneity is conditioned by the three dimensional aspects of the sampling volume. Sampling units should be confined vertically to the soil horizon, or part of the horizon which does not vary significantly with depth for the objective. Horizontally, the sampling units should be confined to an area which can be considered as a unit with regard to soil type, plant growth, or other objectives of the investigation. In many cases, a horizon of a soil type, or a soil phase within a soil type in a given area is considered a sampling volume which is fairly homogeneous.

1.3 TYPES OF SOIL SAMPLING

Sampling soils has often been carried out without due regard to proper procedures, although some information is available in the literature. Cline (1944, 1945), Reed and Rigney (1947), and Peterson and Calvin (1986) have reported sampling procedures, that can serve as guide lines on methods of soil sampling. Sampling is used by business, marketing firms, and polls to investigate large populations. In these instances, sampling is considered a very serious matter and has become an important tool. A number of books are available about the theories and practices of sampling in general (Hansen et al., 1953; Cochran,

1963), and some of the methods can be adapted to soil sampling.

If the soil volume under investigation is small, it is both admissible and convenient to take samples of the whole population. However, with a large soil volume, time and expense can be saved if a representative sample can be drawn and analyzed. The following sections will provide the reader with some general principles as to how representative soil samples can be obtained. These sections will consider several options in cost-effective techniques as well as accuracy and precision in the sampling process.

1.3.1 Simple Random Sampling

Of all the methods in use, perhaps sampling at random is the simplest method. The selection of samples is left completely to the luck of the draw without regard to the variation in a soil population. This is a method by which every sampling unit has an equal, independent probability of being drawn. In a highly homogeneous field, it is often a satisfactory method. However, if the variability in the population is known, one of the following methods to be discussed below may be more adequate.

1.3.2 Systematic Sampling

As the name implies, sampling is done systematically. For example, samples may be drawn at intervals of 5 m, or taken only at footslopes and tops of hills. As a third possibility, the soil underneath every other tree in an orchard may be sampled. This type of sampling can also be combined with simple random sampling as follows. Sampling sites are determined at random on a soil map, and one might draw samples only from sites carrying even (or odd) numbers. In this respect, a lot of other types of systems can be devised for the purpose of soil sampling.

Systematic sampling can often provide more accurate results than simple random sampling, because with this method

Principles of Soil Sampling

the samples are distributed more evenly over the population. However, if the soil population contains a periodic (systematic) variation, and if the interval between successive sampling sites happens to coincide with that of the soil variations, biased samples may be obtained. Therefore, it is advisable to study in advance the nature and occurrence of soil variability before a decision is made to use this method.

1.3.3 Stratified Sampling

This type of sampling is commonly employed in a heterogeneous field. If we can form subpopulations so that a heterogeous population is divided into parts, called strata, each of which is fairly homogenous, more precise sampling can be achieved. The field is thus divided into a number of strata, and a sample or samples are drawn independently in each stratum. The following example serves as an illustration of this type of sampling.

A soil survey indicates that a heterogeneous field can be divided into four fairly homogeneous parts, designated as strata # 1, 2, 3 and 4 with the following distribution:

Stratum # 1 = 1/8 of the field
Stratum # 2 = 1/4 of the field
Stratum # 3 = 1/8 of the field
Stratum # 4 = 1/2 of the field

If 24 samples are to be drawn from the field, then according to the *rule of proportionality* 3, 6, 3, and 12 samples are to be taken from stratum #1, 2, 3 and 4, respectively. The sampling itself in each stratum can be performed randomly or systematically.

1.3.4 Compositing

Compositing is the mixing of sampling units to form a single sample which is used for chemical analysis. This method offers the advantage of increased accuracy through the use of large numbers of sampling units per sample. In compositing, the fundamental assumption is that analysis of the composited sample yields a valid estimate of the mean, which would be obtained by averaging the results of analysis from each of the sampling units contributing to the composite. Cline (1944) indicates that this assumption is valid only if:

1. The sampling volume represents a homogeneous population,
2. Equal amounts of each sampling unit contribute to the subsample analyzed,
3. No changes have taken place in the composite and subsamples prior to analysis that would affect the analytical results, and
4. An unbiased estimate of the mean is the only objective.

Compositing enables the soil scientist to obtain reliable means for a large number of soils at relatively small expense. However, as must be recognized, only the mean is obtained. Hence, no measure of variability is possible, except by analysis of the individual sampling units contributing to the composite. For this reason, compositing, especially vertical compositing for studies of soil genesis, morphology, and classification, is seldom performed. In such studies, variability is frequently of more importance than means. One of the objectives of this type of analysis is the determination of variation in horizons of a particular pedon or between pedons, and compositing would violate the requirement of homogeneity.

However, compositing is applicable to soil fertility studies and soil testing, in which the effect of one or a series of factors is

Principles of Soil Sampling

sought. The result of a given treatment or factor is studied by comparing means for the treated plots with checks.

The common procedure in compositing is to take a number of individual sampling units (slices or cores) according to a sampling method, as discussed earlier. The number of sampling units to make one composite sample ranges from 4 to 16 (Jones, 1985). As a general rule, the total area represented by one composite sample should not exceed 1 ha (100 x 100 m). This limit may be exceeded if the field under observation is reasonably homogeneous in soil characteristics, and if the field can be treated as a single unit in terms of treatments and cropping sequence. In the presence of variations in soil type, sampling units from the different soil types should not be composited, but should be kept separate for analysis.

1.4 MECHANICS OF SAMPLING

The tools that can be used to collect the sampling units include soil sampling tubes, augers, and till spades or shovels, depending on the depth and types of mechanics of sampling. As indicated earlier, samples can be dug from the soil surface (shallow sampling); or from deeper layers (deep sampling); or from trenches, road banks, and soil pits.

1.4.1 Shallow Sampling

__Undisturbed__ Samples are collected with a box (die) or a tube, which is driven carefully into the soil surface. The sample is cut loose with a knife. Alternatively, in well aggregated soils, undisturbed soil aggregates can be cut loose with a knife.

Undisturbed samples are needed for bulk density measurements and soil fabric or thin section analysis.

__Disturbed__ Samples are collected with either an auger, a tube or a core, or with a shovel or a spade. When a spade or a shovel is used, the sample is collected by cutting a V-shaped slice to the proper depth. A knife is then used to cut and lift the center portion of the sample, which is placed in a clean plastic bag for transport to the laboratory. If a composite is to be drawn, the collected units are put into a clean plastic bucket and mixed thoroughly until a completely homogeneous mixture is attained. A small portion adequate for analysis is then transferred to a plastic soil bag for transfer to the laboratory.

1.4.2 Deep Sampling

Deep sampling can be done by using tubes or cores, augers, or spades, or by utilizing trenches and road banks, or by digging soil pits and exposing soil profiles. The use of augers and coring devices may increase the chances of contamination, and sampling road banks or digging soil profiles is preferred.

1.4.3 Soil Profile Sampling

Soil profile sampling is usually made for (1) physico-chemical, and mineralogical analysis of importance in soil genesis, morphology, and classification, and for (2) chemical analysis needed for soil testing, or studies in soil fertility.

Plant roots penetrate different horizons, and each horizon provides a different chemical environment for the growing roots. To reduce time and expense in the analysis, the root zone in soils is commonly sampled. Frequently, this sample is composed of the A and part of of the B horizons. Although this procedure violates the requirement of homogeneity, a root zone can be defined as one unit for the objective.

In the study of soil genesis, morphology, and classification, soil types are usually selected to represent the soil volume. The use of such representative samples simplifies sampling procedures, analysis, and interpretation of results. However, soil types may vary from one another, and their properties may

Principles of Soil Sampling

accordingly differ, depending upon the factors of soil formation. To isolate the effect of a single factor, soil types can be sampled according to any of five criteria:

1. *Climosequence*, which may produce results on the effect of differences in climate.
2. *Lithosequence*, which will provide information on the effect of differences in parent material or geologic formations.
3. *Biosequence*, which will show the effect of differences in vegetation, e.g., differences between grassland and forest soils on similar parent material.
4. *Toposequence*, which will provide information on the effect of differences in topography or relief.
5. *Chronosequence*, which yields information on the effect of time and age in soil formation.

However, it must be noted that variation in one soil formation factor may affect in turn changes in the other factors. For instance, climatic changes usually have a significant effect on vegetational changes.

1.4.4 Time of Sampling

Normally, sampling instructions do not specify a particular time of sampling, although seasonal cycles in soil test parameters are present. Generally the best time to take samples is in midsummer to early fall, when seasonal effects are minimal. Some workers recommend taking soil samples at the same time plant tissue is being collected for analysis, normally during mid- or late-summer. However, the actual time for soil sampling may be dictated more by the cropping sequence than by the season of the year. For the purpose of soil genesis, morphology and classification, no particular time is required for sampling. Sampling for pedology usually depends upon conditions in the field. Sampling is seldom conducted in winter, when the soil is hard and frozen solid, or during rainy weather in spring, when the soil is muddy and puddled.

1.5 SIZE AND ACCURACY OF SAMPLING

The size of sampling can be distinguished into (1) size (quantity) of samples and (2) number of samples to be taken.

The size of the sample or the size of the sampling unit to be drawn depends upon three factors: (1) coarseness of the material, (2) objective of the analysis, and (3) the desired accuracy. Little information on this subject is available in the soils literature. Perhaps the information in Table 1 illustrates the problem. As can be noticed from the data, the coarser the material, the larger the sample size required.

On the other hand, the number of samples to be taken is determined principally by the objective and the cost. The more heterogeneous the soil population, the more intense must be the sampling rate to attain a given accuracy. The greater the number of samples and the smaller the sampling volume, the more precise will be the estimate of the soil characteristic analyzed. Although frequently the number of samples is decided arbitrarily, economic considerations often restrict both the quantity and the number of samples taken. In such cases, a suitable balance must be found between the amount and number of samples to be taken as required for accuracy, and the amount and number as desired by economy. The question is now how many samples should be taken and analyzed in order to make the error as small as possible. The objective is to avoid taking too many samples, while on the other hand we must also avoid taking too small a number of samples so that the results can become too inaccurate. Although a precise answer to the problem is not easily determined, a statistical approach has been developed (Peterson and Calvin, 1986; Snedecor and Cochran,

Principles of Soil Sampling

Table 1.1 Quantity of Sample in Relation to Coarseness of Material (Krumbein and Pettijohn, 1938)

Material	Diameter (mm)	Suggested wt of sample	Approx. vol of sample
Cobbles	128 - 64	32 kg	---
Pebbles	64 - 4	16 - 2 kg	---
Gravel	4 - 2	1 kg	1000 mL
Sand	2 - 0.050 mm	500 - 125 g	500 mL
Silt	0.05 - 0.002 mm	125 g	250 mL
Clay	< 0.002 mm	125 g	250 mL

1974). First, a decision must be made on the magnitude of the error that can be tolerated, called the acceptable error = E. This error can be calculated by:

$$E = \pm\, t\, (V)^{0.5} \tag{1.1}$$

$$V = S^2/n \tag{1.2}$$

in which t = t-test value, V = variance, S^2 = sum of squares, and n = number of treatments or in this case number of samples.

The required number of samples dictated by the acceptable

error can then be calculated by using one of the following formulas:

$$n = 4\sigma/E^2 \tag{1.3}$$

or

$$n = t^2 S^2/E^2 \tag{1.4}$$

in which n = number of samples, and σ = standard deviation.

Equation (1.4) is preferable, since it takes into consideration the probability level by including the t-test. The standard deviation is calculated by the usual formula:

$$\sigma = \sqrt{\Sigma(x_i - \bar{x})^2/(n - 1)} \tag{1.5}$$

whereas the sum of squares, S^2, is found by using the equation:

$$S^2 = \Sigma(x_i - \bar{x})^2/(n - 1) \tag{1.6}$$

The above formulas can be used for determining the number of samples that should be taken in random, systematic or stratified sampling plans, according to limits set for acceptable error. The following two examples serve as illustrations:

Principles of Soil Sampling

Example A Assume that 10 samples were taken at random from 0-10 cm depth in an experimental field. Analysis of the 10 samples for Ca concentration yields the following results:

Sample #	1	2	3	4	5	6	7	8	9	10
Ca (x_i)	90	94	100	79	97	104	94	88	84	90 ppm
$\bar{x}$ (mean)	92	92	92	92	92	92	92	92	92	92
($x_i - \bar{x}$)	-2	2	8	-13	5	12	2	-4	-8	-2
($x_i - \bar{x}$)^2	4	4	64	169	25	144	4	16	64	4

$\Sigma(x_i - \bar{x})^2 = 498$

$S^2 = 498/9 = 55.3$

V (variance) = 55.3/10 = 5.53

E (10% level) = ± 2.26 (5.533)$^{0.5}$ = ± 5.32 ppm

The above result means that in taking 10 samples, an error of 5.32 was tolerated, and the true mean value of the Ca content lies between 86.68 and 97.32 (92 ± 5.32).

If this level of accuracy is not acceptable and an error of 2 ppm is wanted, then the number of samples required with an acceptable error of 2 ppm is:

n = (1.83)^2(55.33)/2^2 = 46.3

Example B Stratified sampling and analysis of the individual samples for K content yield the following results:

Stratum a: 60 40 ppm K $(x_i - \bar{x})^2 = 200$
Stratum b: 56 49 51 47 ppm K $(x_i - \bar{x})^2 = 44.75$
Stratum c: 70 62 59 53 68 65 ppm K $(x_i - \bar{x})^2 = \overline{194.86}$
                                    $\Sigma(x_i - \bar{x})^2 = 439.61$

The data above indicate that:

n = 12, and degrees of freedom (d.f.= n - 3) = 9

Therefore: S^2 = 439.6/9 = 48.84

V = 48.84/12 = 4.07

E = $\pm\, t_{0.1}\, (4.07)^{0.5}$ = ± 3.67

The true mean value lies between 52.97 and 60.35 (56.66 ± 3.67). If the error above is nonacceptable, and we want to decrease the error, E, to 2.00 ppm (with a 5% level probability), the number of samples (n) is then:

n = $(2.26)^2(48.84)/2^2$ = 62.4

This could mean that the number of samples is too large to be economical from the standpoint of time required for analysis and the cost incurred.

CHAPTER 2

SAMPLE PREPARATION

2.1 SAMPLE PRESERVATION

After the gross sample has been collected, care must be taken to avoid contamination, and to prevent the occurrence of further chemical reactions. The sample must be preserved as much as possible in its original condition, and it is necessary to maintain the properties and the identity of the sample at all stages of sample preparation. No method should be used in the preparation of the sample that can either alter or destroy the information to be obtained in the analysis.

Air drying is the most accepted procedure of sample preservation. Drying in the air may reduce the rate of possible reactions in the disturbed soil sample. Additionally, a sample should not be allowed to stay moist for extended periods of time. Soil aggregates should be broken carefully to accelerate the drying procedure. It is generally assumed that chemical and biochemical reactions in air-dry soils are reduced to a minimum, although these reactions are still possible sources for errors.

To accelerate the drying process, samples may be placed in a forced draft of moving air, but not in heated air. The temperature must not exceed 35°C because drying at elevated temperatures may cause drastic changes in the physical and chemical characteristics of the soil sample (Hesse, 1972). Drying may also result in increased cementation, which may particularly affect particle size distribution analysis. Other possible results from drying at high temperatures are chemical changes in the oxidation status of elements and microbiological reactions. The degree to which such changes occur varies with temperature and time of drying. Some of the physical and chemical effects of drying soil samples are discussed below in more detail.

Particle Size Distribution Drying at elevated temperatures has been found to affect results of particle size distribution analysis (Van Schuylenborgh, 1954). As indicated above, the drying process increases cementation of soil aggregates, hence producing only a partial dispersion during the analysis. The latter results in a particle size distribution that indicates a sandier (lighter) texture than is usually produced. However, authorities have not reached uniform agreement on the relative merit of performing particle size distribution analysis on samples in their natural moist state.

Soil Moisture Content Use of air-dry soil is recommended for soil moisture content determination. Most of the soil moisture has been evaporated by drying at high temperatures, hence there is no use in measuring the amount of moisture in the sample. An oven-dry soil also tends to become hygroscopic and will adsorb moisture from the air. It is virtually impossible to prevent oven-dry soil samples from adsorbing moisture during handling. Errors in weighing these soils are considerable, but no serious errors results from the use of air-dry soil.

Organic Carbon Air drying will not affect total carbon analysis, but oven drying may cause some C to be lost due to oxidation of organic matter. In microbial studies, drying at elevated temperatures may also cause destruction of a number of microorganisms.

Nitrogen Air drying soil samples prior to analysis has little effect on total N analysis. However, it has been found that in the determination of ammonium- and nitrate-N, these N fractions increase with time and temperature of drying. This effect is greatly enhanced if the sample is dried at elevated temperature in the oven. Such a drying procedure may affect the ammonifying and nitrifying organisms. Air drying the samples will have little effect upon the nitrifying organisms, but re-wetting the sample results in a renewal of microbial activity at an increased rate. The latter may cause the nitrate content of the sample to increase. On the other hand, drying at elevated temperatures would destroy the nitrifyers but not the ammonifyers. Consequently, re-wetting of the sample may cause ammonium-N to accumulate. When analyses for NH_3- and NO_3-N are to be made, one suggested method is to make immediate analyses on the moist sample.

Soil pH The effect of air-drying on soil pH measurements and especially drying soil samples at elevated temperatures has been studied in a number of investigations. Drying usually results in a decrease in soil pH, the effect being both chemical and biological. The reduction in soil pH, as a result of drying soil samples at elevated temperatures, is especially pronounced in soils rich in sulphur. Upon drying at elevated temperature, S tends to be converted into its oxidized forms. Sulphur dioxide may dissolve in soil water to produce sulphuric acid, causing the decrease in soil pH (Van Schuylenborgh, 1954).

Phosphorus Drying can cause changes in P fixation, which is related to changes in Al and Fe chemistry. Drying of soils, particularly at high temperatures, may sometimes increase dilute-acid soluble P, which perhaps occurs in conjuction with a decrease in the soil pH, as discussed above.

Sulphur Drying soils at room or elevated temperatures may cause the S to oxidize. The SO_2 formed dissolves in water to form sulphuric acid. Not only will this affect changes in soil reaction as discussed above, but will also tend to increase the sulphate content in the soil solution.

Potassium During the drying of soil samples, K may be fixed or released from the fixed form. The reaction that occurs depends upon the level of exchangeable K and on the type of clay minerals present. Therefore, the results of K determination may be affected considerably by the method of drying the samples.

Iron When soil samples are dried, especially at elevated temperatures, all the iron is oxidized into Fe(III), the most stable form of Fe. Determination of Fe(II) is then made impossible.

Manganese Drying of soils increases the exchangeable Mn content. As reported by Sherman and Harmer (1943), freshly collected samples are needed for determination of exchangeable Mn. Several authors (Fujimoto and Sherman, 1945, 1948) believe that the drying effect is related to a reduced biological activity, e.g., reduced bacterial oxidation of Mn to the oxide form. This results in more manganese remaining in the ionic form, which then can be adsorbed by the clay complex.

2.2 SAMPLE STORAGE

Finally worth mentioning is storage of samples. Contamination of samples because of adsorption of gases is a frequently overlooked source of error. Samples should be dried and stored in air-tight containers as soon as possible to avoid adsorption of NH_3, SO_3 and/or SO_2 gases in the laboratory. Containers should be clean and should be composed of materials that will not contaminate the sample. Glass jars, or plastic or waxed card-

Sample Preparation

board containers are suitable for this purpose.

2.3 SAMPLE PREPARATION

As indicated before, the sample is composed of many individuals, which form a population of soil constituents of widely varying size and composition. The inorganic soil fraction varies in size depending upon whether gravel, sand, silt, or clay constitutes the sample. The size distribution of the soil minerals can also differ widely. For instance, quartz is often concentrated in the coarse particle fractions, whereas clay minerals are accumulated in the fine particle fractions. In well-aggregated soils, the inorganic soil constituents are present with the organic soil constituents in the form of aggregates. Aggregation tends to contribute greatly to heterogeneity within a given fraction. The whole heterogenous mixture must be prepared and subsampled for analysis in such a way that the analysis will yield results that identify the soil property accurately. This objective requires that during subsampling every particle has an equal chance of being drawn by thorough mixing.

2.4 GRINDING AND SIEVING

2.4.1 Grinding

Thorough mixing requires that the sample be crushed and ground to particles of uniform size. Large aggregates are reduced by crushers, and the crushed sample is then further reduced by grinding. Care should be taken not to break the individual soil minerals during the process. The purpose of grinding is to reduce heterogeneity and to provide maximum surface area for physical and chemical reactions. Various devices are used for crushing and grinding soils. The choice of equipment depends on four factors:

1. The amount of sample to be crushed or ground
2. The degree of fineness to be attained
3. The contamination that can be tolerated
4. The analysis in question

Crushing soil aggregates can be done by using jaw crushers, hardened steel mortars or rocking boards. Jaw crushers are used for crushing large aggregates, steel mortars for smaller aggregates, and rocking boards for small aggregates. The types of grinders used are plate grinders (friction mills), ball mills, rod mills, agate mortars, and boron carbide mortars. For large amounts of samples, plate grinders or ball or rod mills are used. The sample is mixed with stainless steel rods or balls in a container and the container is rolled over a period of time. When properly operated, rod and ball mills are efficient grinders (Coghill and Devaney, 1937). For small amounts of samples, an agate or boron carbide mortar can be used.

2.4.2 Sieving

Sieving is an essential part of homogenizing the sample after the grinding operation. All particles will not be pulverized in a single grinding operation because some of the very hard particles become coated with powdered material. The fine material cushions and prevents these hard particles from being thoroughly ground. The bulk sample is then screened, and the rejected material ground again, and rescreened until the whole sample passes through the sieve.

In the past, a 0.5-mm sieve was used for this purpose (Assoc. Official Agric. Chem., 1940), but currently a 2-mm sieve is preferred (Robinson, 1930; Jackson, 1958; Soil Survey Staff, 1972). The fraction passing a 2-mm sieve is collected and stored as a stock sample. The procedure of passing materials through a 2-mm sieve is consistent with the internationally accepted standard for chemical analysis. Analysis of samples that are passed through sieves of other sizes may produce results that

Sample Preparation

would not be comparable to accumulated data in the world (Cline, 1944).

It is also believed that the soil fraction <2 mm represents more accurately the total soil volume than do the finer sieve size fractions. However, sieves finer than 2 mm are, of course, also employed. The different sand fractions are obtained by sieving through sieves of finer sizes, and where microanalysis requires the use of a very small amount of sample, the stock sample (<2 mm fraction) can be subsampled, ground, and screened through a 1-mm or 0.5-mm sieve. The fraction <0.5 mm, preferred in modern microanalysis, provides the needed margin of safety. Ordinarily, an extremely fine sample is not essential in soil analysis, and efforts spent in achieving such a fine sample is needless and may even be harmful.

The soil fraction >2 mm is usually discarded in soil chemical analysis. These materials are not soil constituents, but are rock fragments which may produce soil constituents after weathering. However, in studies of soil genesis, morphology, and classification, this fraction is examined for the presence of concretions and gravel content. Concretions may be collected and analyzed further, depending upon the objective of the research.

Results obtained by analyses of samples passing different sieves are not necessarily comparable. Resistant soil minerals are often accumulated in the coarse size fraction and are discarded by sieving with finer sieves. Coarse particles are also discarded by using finer sieves, causing the analyzed sample to be biased in favor of fine particles.

2.4.3 Errors in Grinding and Sieving

Contamination by Grinding During the process of crushing and grinding, some contamination from the apparatus occurs due to abrasion with the soil sample. Crushing generally results in less abrasion of the apparatus than does grinding. Hardened steel grinders are very susceptible to abrasion, even by soil minerals that are less hard than quartz. Porcelain mortars and ball mills are unsuited for grinding materials for quantitative analysis. Agate mortars have been found more suitable for this purpose.

Vertical blows may crush quartz sufficiently fine for further grinding in an agate mortar. Boron carbide mortars are slightly less hard than agate mortars.

Contamination in Sieving Brass and copper screens are usually employed in sieving soils because of their resistance to corrosion and rusting. However, some abrasion of the brass or Cu during sieving may take place, causing the sample to become contaminated with the metal. Because this kind of contamination is seldom serious in routine macroanalysis, brass and Cu sieves are usually satisfactory. However, in an analysis of trace elements, their use would be the reason for some errors. In such an analysis, silk bolting cloth may be used for screening.

Oxidation of the Sample Considerable heat is generated locally during grinding, and in the presence of air an appreciable amount of oxidation of soil constituents may occur. This possibly is particularly critical in soils containing ferrous compounds, which are quickly oxidized into the ferric state.

Loss or Gain in Moisture The heat generated during grinding may result in loss of water from the sample, especially when grinding is continued over a period of time. In the latter case, the crystal structure, which is very important in clay mineral analysis, may even be destroyed.

Many materials, not normally hygroscopic, become rapidly hygroscopic when converted into a finely divided state by grinding. The fine powdered material may adsorb moisture from the air and present difficulties in handling and weighing.

Sample Preparation

2.5 REDUCTION IN SAMPLE SIZE

When a very large sample is taken and processed, it may be necessary to reduce the amount of the sample for ease of storage and handling. The mixture obtained by grinding and sieving must be reduced in size and subsampled for analysis in such a way that a single analysis can properly identify the soil characteristic under investigation. This objective requires that

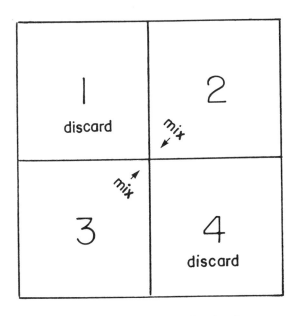

Figure 2.1 Reducing sample size by *quartering*.

every particle must have an equal chance to be subsampled. In order to achieve this result, a random and unbiased method of subsampling is essential. Reduction in size or *sample splitting* can be performed with a mechanical sample splitter, such as a riffle sampler, by which the sample is automatically divided in half by a series of chutes (Krumbein and Pettijohn, 1938; Willard and Dhiel, 1943; Hesse, 1972). This process can be repeated as many times as necessary. Manually, the sample is usually reduced in size by *quartering*. The sample is spread uniformly over a sheet of paper or plastic and divided into four equal portions (Figure 2.1). Portions 2 and 3 are collected and thoroughly mixed, whereas the remainder is discarded. This process of quartering can be repeated as many times as necessary until the right size of sample is attained. In case the sample is heterogeneous in particle size, the different materials tend to separate. Micaceous materials also have a tendency to separate easily. Bailey (1909) believes that a riffle sampler is more efficient than quartering for reducing the size of the sample.

CHAPTER 3

ERRORS AND VARIABILITY OF SOIL ANALYSIS

3.1 MEASURE OF ACCURACY

A soil chemical analysis is often very complex, and the result of analysis may not always be reliable. The analysis can be affected by many errors, some physical and some chemical in nature; thus it is by no means easy to obtain satisfactory results in all cases. Errors affecting the results should be detected and assessed, and the analysis repeated if necessary. Error affects the *accuracy* and *precision* of the analysis. The latter two terms are frequently used interchangeably. By definition, an error (E) is the difference between a value obtained by analysis (X) and the true value (T) of the soil characteristic:

$$E = X - T \qquad (3.1)$$

The true value can be distinguished into (1) a statistical true value or operational true value, which equals the arithmetical mean of results of an infinitely large number of analysis; and (2) the scientific true value, which is the real value, e.g., as conditioned by formula weight.

If the real true value can be determined, the magnitude of deviation from the true value, called *error*, is a measure of accuracy of the analysis. The smaller the difference, the more accurate is the analytical value. This type of error is seldom expressed in terms of absolute values, but is often calculated in percentages with respect to the true value, T. This is called the *relative error*.

Example A technician analyzes the Na content of NaCl and determines that the Na content is 35.0%.

Question Calculate the absolute and relative error. The atomic weight of Na = 22.99, and that of Cl = 35.46. The molecular weight of NaCl = 58.45.

Answer The scientific true value of the Na content equals:

$T = 22.99/58.45 \times 100 = 39.3\%$

Therefore, the absolute error equals:

$E = 35.0 - 39.3 = -4.3\%$

The relative error = $-4.3/39.3 \times 100 = -10.9\%$

3.2 MEASURE OF PRECISION

In many cases the scientific true value of a soil property can not be determined. The error must then be evaluated without employing the true value. The latter is performed by calculating

Errors and Variability of Soil Analysis

the difference between the analytical value obtained and the mean value, M (= the average value of a number of measurements). This difference is called the *deviation* (D), which can be formulated as follows:

$$D = X - M \qquad (3.2)$$

The magnitude of this deviation is a measure of precision of a result of an analysis. It is seldom expressed in absolute values, but is often presented in percentages with respect to M. The smaller the value of D, the more precise the result is to M. Thus, it can be inferred that by increasing the number of analysis to an infinitely large number, the mean value, M, will equal the arithmetical true value. The latter does not necessarily equal the scientific true value, which will be discussed in the section on bias. For this reason, the two terms accuracy and precision should not be used interchangeably. *Accuracy* reflects the correctness of the value with respect to the scientific true value, whereas *precision* only indicates the closeness of the value to the mean; hence precision reflects the reproducibility of the analysis. Precision can be increased by greatly increasing the replications of analysis (Kempthorne and Allmaras, 1965).

Example Cation-exchange capacity determination of a soil yields the following results:

		CEC (cmol(+)/kg)
Determination	#1	20.0
"	#2	22.0
"	#3	17.0

Question Calculate the relative deviation of result #1.

Answer The mean value of the three CEC determinations = 19.67 cmol(+)/kg.

Therefore, the absolute deviation of result #1 = 20 - 19.67 = 0.33 cmol(+)/kg.

Hence, the relative deviation equals: 0.33/19.67 x 100% = 1.7%

Similar calculations of absolute deviations can be made for determinations #2 and #3. A series of individual deviations from the arithmetical mean is then obtained, which serves as the basis for the formulation of *standard deviation*. The latter, also known as the *standard error* (Snedecor and Cochran, 1974), can be calculated with equation (1.5). The standard deviation, is therefore an extension of absolute deviation, D, and is frequently employed because it is impractical to perform an infinitely large number of analyses.

3.3 MEASURE OF BIAS

Bias is defined as the difference between the statistical mean and the scientific true value (Allmares, 1965). In an analogy to the above errors, bias can consequently be formulated as follows:

$$B = M - T \qquad (3.3)$$

in which B = bias, M = statistical mean, and T = scientific true value. The statistical mean is the arithmetical mean of a number of analyses. If M is calculated from a large number of results, it equals the statistical true value or operational true

Errors and Variability of Soil Analysis

value, which has been defined earlier.

If the difference is attributed to improper methods or scientific approach, it is called a *scientific bias*. On the other hand, if the difference is the result of analyzing non-representative samples, it is called *sampling bias*. If the instrument or chemicals employed yields improper measurements, this type of bias is called *measurement bias*. Many other types of bias can be formulated in this way.

From the above, it can be noted that bias is just another type of error. It can be distinguished from accuracy and precision, that bias is the difference between the mean value and the scientific true value as spelled out in the definition, hence it reflects both accuracy and precision. If the result is not biased, then it is highly accurate and precise or reproducible.

Bias can be in the form of an addition to or a subtraction from the scientific true value, which is then called *additive bias*. Additive bias is formulated as follows:

$$M = k + T \qquad (3.4)$$

in which M = biased statistical mean, and k = error, or the difference between M and T. Hence the error k equals B = bias. According to the definition of bias, the value of k can, therefore, be negative or positive in sign.

Bias can also manifest itself as the error k multiplied with the scientific true value:

$$M = kT \qquad (3.5)$$

This is then known as a *multiplicative bias*. In this case the value of k is positive in sign.

3.4 TYPES OF ERRORS

Two main groups of errors are usually recognized: (1) errors that can be determined, known as *determinate errors*, and errors that cannot be determined or that are very difficult to determine, termed *indeterminate errors*.

3.4.1 Determinate Errors

Because this group of errors can be determined, their influence on the results can be determined. The determinate errors that can affect a soil chemical analysis are numerous and only a few examples of these errors will discussed below.

Instrumental Errors These are determinate errors and can be traced to the instruments and chemical reagents used in the analysis. The use of uncalibrated balances may yield errors in weighing. Since this kind of error remains constant in value in repeated weighing procedures, it is often called a *constant error*. Another group of instrumental errors results from the use of uncalibrated glassware, e.g., volumetric flasks. Errors may be caused by expansion and contraction of volumetric solutions, and the erroneous values will change systematically, depending upon changes in temperature. Since the expansion and contraction coefficient of the glassware is known, the error can be calculated. This kind of determinate error is called a *systematic error*. Another type of systematic error is introduced by impurities in the chemical reagents used.

Operative Errors These errors are not caused by the instrument, the glassware or the chemical reagents, but rather are caused by the operator who performs the analysis. Operative errors can be significant in value and an inexperienced, careless, or thoughtless worker can even completely destroy the analysis.

Errors and Variability of Soil Analysis 33

Using nonrepresentative samples, spilling liquids during measurements, allowing uncontrolled frothing during a boiling procedure, underwashing and overwashing precipitates, and introducing foreign material into samples are examples of operative errors.

Personal Errors These are the errors often related to judgments made by persons performing the analysis. They may be caused either intentionally or unintentionally. Two types of personal errors can be distinguished, personal judgment, and prejudice.

A *personal judgment error* is unintentional. For example, in performing a volumetric analysis, some workers are unable to determine the exact color change that indicates the end point of titration. They will always *overshoot* or *undershoot* the end point. For one operator, this error can be more or less constant in magnitude. Fortunately, this error can be eliminated by using potentiometric titrations for the detection of the end point.

Prejudice, a more serious form of personal error, may be manifested when analyses are replicated. Prejudice errors occur when the operator tends to choose in duplicate analyses the results that agree most closely with previous findings. In replicated volumetric analyses, the operator tends to stop the titrations at the points which will give results in agreement with one another. This type of error may also occur when the operator is reading the pH scale in pH measurements or calculating d-spacings in x-ray diffraction analysis. When there is some doubt as to the exact place on the division in reading the scale for the above measurements, the operator is likely to choose the reading that agrees closely to the first one. Even conscientious investigators can be influenced by personal prejudice, and others, less conscionable, are too willing to avail themselves of an excuse to make results in the replications agree with one another.

Analytical Errors These errors are attributed to analytical procedures and methods, and they are related to the physicochemical characteristics of the system. They are sometimes referred to as *errors of methods*, as they can be traced to the procedures, and not to the operator. Regardless of the care and skill of the worker, an analytical error is uncorrected unless the procedure itself is changed. An example of this kind of error, is the error introduced by the washing procedure of a cation-saturated soil in cation-exchange capacity determination. Other examples are side reactions or interactions, co-precipitation, and failure of a reaction to go to completion during analyses.

3.4.2 Indeterminate Errors

Indeterminate errors, also known as *accidental errors*, cannot be determined. For example, the results of repeated analyses performed by the same person under similar constant conditions may differ, and the reasons for and the magnitude of changes are unknown. It is believed that the operator cannot prevent these errors from occurring. They tend to occur at random and their occurrence, therefore, obeys the probability law of a normal distribution. The latter, represented by the famous *bell* curve, is often formulated in different fashions. One of the simplest equation is as follows (Snedecor and Cochran, 1974):

$$Y = h/\sqrt{\pi} \, \exp.\{-h^2(x - \bar{x})^2\} \qquad (3.6)$$

in which h = frequency of occurrence of the error, x = variable, or analytical result, $\bar{x}$ = mean, e = 2.71183, and π = 3.1416. Many authors consider h = 1, but Snedecor and Cochran (1974) indicate that h = $1/\sigma\sqrt{2}$.

Errors and Variability of Soil Analysis

3.5 REJECTION OF RESULTS

The reliability of results is often a matter of opinion, and various authors give various criteria for the rejection of results. In general, however, the results are reliable, if they have high accuracy and precision and are without bias. Many methods can be used to check accuracy, precision, and bias, and frequently the same method is appropriate for testing accuracy as well as for precision and bias.

3.5.1 Checking for Accuracy

As stated before, accuracy expresses the correctness of a result with respect to the scientific true value. Although many methods are available to check for accuracy, the following three major methods will be discussed in more detail. They are the method of summation, the method of ionic balance, and the use of independent methods.

Method of Summation This method finds application especially in elemental analyses of soil minerals and rocks, and in particle size distribution (soil texture) determination. A fundamental assumption in this method is that summation of the results of a perfect analysis will yield 100%. This theoretical value is seldom obtained, depending upon the error. In particle size distribution analysis by the pipette method, summation of the sand, silt, and clay fractions may yield a value which is higher or lower than 100%. The acceptable error is subject to many questions, since the error depends upon many factors. In the elemental analysis of rocks and minerals, the error depends also on the number of soil constituents analyzed. The smaller the number of constituents contributing to the summation method, the larger will be the deviation from 100%. However, a proper analysis should yield in all cases a summation which approaches 100%. In the elemental analysis of soil minerals and rocks, a summation value between 99.75 and 100.25% is acceptable. The following data serve as an example:

Table 3.1 Elemental Composition of Basalt and Andesite Rocks (Mohr and Van Baren, 1960)

Composition	Basalt (%)	Andesite (%)
SiO_2	50.78	52.89
Al_2O_3	16.15	18.93
Fe_2O_3	2.47	4.06
FeO	6.90	4.80
TiO_2	1.45	0.96
MnO	0.12	0.43
MgO	8.40	3.72
CaO	8.33	8.56
K_2O	1.18	0.94
Na_2O	3.45	3.98
H_2O	0.56	0.58
Total	99.79	99.85

These results are therefore acceptable.

The Ionic Balance In soil analyses the operator frequently deals with a system that is composed of cations (positively charged particles) and anions (negatively charged particles). To maintain electro-neutrality, these ions exist in the soil solution in equivalent amounts. This principle is used for testing the accuracy of results in soil analysis involving ions. The sum of cations must, therefore, equal the sum of anions, expressed in

Table 3.2 Cation-Anion Balance in Soils

cations	me/100g	anions	me/100g
Ca^{2+}	203.6		
Mg^{2+}	110.1		
Na^+	070.0		
K^+	015.0	Cl^-	106.1
Al^{3+}	010.0	SO_4^{2-}	208.0
H^+	010.0	HCO_3^-	103.6
Total	418.7	Total	417.7

terms of equivalents or milliequivalents per unit amount of soil. Table 3.2 serves as an example.

It can be noticed that the difference between cation and anion concentrations (in milliequivalents) is acceptable, since it differs only 0.24%. Theoretically the me of cations must equal the me of anions, but a perfect match between me-cations and me-anions in a natural soil solution is very rare. The error becomes larger or smaller, depending upon the number of ions analyzed that contributes to the summation process. This method is of special importance when the scientific true value is unknown.

Independent Methods Independent methods are used to check accuracy in the absence of a scientific true value. Frequently, analyses can be performed by two completely different methods. For example, the determination of Al in soil solutions can be conducted by colorimetric methods or atomic absorption spectroscopy. If the results of the two completely different methods produce substantially similar results, it is highly probable that the analytical value obtained is correct within only a small range of error.

3.5.2 Checking for Precision

As defined earlier, the degree of precision is measured by the deviation of results from a mean value. Close agreement with the mean demonstrates only that the determination is reproducible, and does not necessarily suggest accuracy of results. Two methods for checking the precision of results are available: statistical analysis and parallel determination.

Statistical Analysis This is perhaps the most rapid and efficient method to check for precision, and several statistical methods can be applied. A statistical measure of precision often used is the *standard deviation*, or *standard error*, σ, which can be calculated by using equation (1.5) listed in the preceding pages. If the number of analysis is very large, equation (1.5) can be changed into

$$\sigma = \sqrt{\{\Sigma(x - \bar{x})^2\}/n} \qquad (3.7)$$

With a large number of n (=number of results), the figure 1 in (n-1) can be neglected, resulting in equation (3.7). A variation of this method is the average deviation, which is formulated as follows:

$$\delta = \Sigma(x - \bar{x})/n \qquad (3.8)$$

Another method frequently used in checking variations between replicated analysis is the coefficient of variations, known as C.V.:

$$C.V. = \sigma/\mu \qquad (3.9)$$

Errors and Variability of Soil Analysis 39

in which σ = standard deviation and μ = mean of results.

The limiting values for σ, and C.V. usually depend upon how large an error the operator is willing to accept.

Parallel Determination Parallel determination refers to running analyses in replications, at least in duplicates. The results of two or more analyses should agree reasonably well with each other, although such an agreement does not guarantee the accuracy of results with respect to the scientific true value. Satisfactory agreement between results only indicates that the analyses are reproducible. This fact often increases our confidence in the reliability of the analysis. The limit for rejection depends again on the magnitude of an error that can be tolerated.

A variation of this method is *independent parallel analysis*, in which a parallel determination is carried out by two different persons. Agreement in results not only suggests reproducibility, but also signifies accuracy of analysis.

3.5.3 Checking for Bias

Since bias refers to errors related to both accuracy and precision, the methods reported for testing accuracy and/or precision often apply equally well for testing the presence or absence of bias. For example, the use of independent methods, summation, the ionic balance, and independent parallel analysis are also considered appropriate tests for checking bias. Other tests that can be used to determine the presence or absence of bias are the methods of *standard addition* and *internal consistency* (Allmaras, 1965).

Method of Standard Addition The principle in the test is to carry out the analysis by adding or deleting the constituent under investigation. Bias is present if the results do not obey the linear relationship:

$$Y = a + bX \qquad (3.10)$$

where Y = observed value, X = concentration of constituent added or deleted, a = intercept with Y axis (≥ 0), and b = regression coefficient or slope of the curve.

This method is easily applied in the analysis of chemical solutions in which chemical constituents can be readily deleted from the solution. However, it is often difficult to delete a constituent from the soil sample. An easier way is to increase the concentration by adding the constituent under observation.

Method of Internal Consistency In this method, the results of an analysis are tested by increasing or decreasing the amounts of soil. The assumption is made that the constituent measured should increase or decrease proportionally to the amount of soil that is added or reduced. Bias is present if the results do not obey the linear relationship as formulated by equation (3.10).

3.6 STATEMENT OF RESULTS

The result of an analysis must be presented in such a way that the accuracy and precision of analysis are properly reflected. The number of figures used in the final value should show the degree of accuracy and precision with which the analysis was carried out. This number of figures, called *significant figures*, is composed of all the digits that have significant or practical meanings for the objective. The significant figures in a final value should include all the certain digits and only one doubtful digit. If, for example, a soil sample must be weighed to the nearest 0.1 mg, it is proper to record the weight to the fourth decimal place, e.g., 10.1234 g. This figure contains six significant digits; the first five are certain digits and the last or the sixth digit is the doubtful digit. Thus, a statement that the weight is 10.123 g is considered incorrect, since this implies that weighing is done to the nearest 1 mg.

Care must be exercised not to decrease the accuracy of the analytical result by improper approximations during calculations. It is always considered best to add one more digit beyond the last significant figure in order not to alter the final value during calculations. For example, the average of 50.05, 50.07, and 50.07 is calculated to be 50.063. The individual three figures, each composed of four digits, are considered significant figures. It is, therefore, correct to state the average = 50.06 for a final statement. This indicates that accuracy and precision of analysis were carried out to the nearest fourth digit. However, when further calculations are needed, the figure 50.063 should be used so that the significant figure (four digits) will not be altered by unwarranted approximations.

The agreement between replicates should also be considered in the decision as to the number of digits to be used in the final value. If duplicate analyses yield the figures 100.04 and 100.59, the two last digits are then uncertain values. Hence, it is correct to record 100.3 as the average. If, on the other hand, duplicate analyses produce the figures 100.02 and 100.05, only the last digits are doubtful. The average is calculated to be 100.035, but this figure should be recorded as 100.04, because the latter shows that the figure is correct to the fourth digit, whereas the last digit is the doubtful digit. As noticed, in rounding off, the rule is to increase the significant figure with 1 if the succeeding digit is 5 or greater, although in computer calculations this last digit is just dropped if it equals 5.

CHAPTER 4

METHODS OF SOIL CHEMICAL ANALYSIS

4.1 PURPOSE AND USE

Two basic concepts are commonly accepted in the scientific study of soils. The first considers the soil as a natural habitat for plants and justifies soil analyses primarily on this basis. This concept, known as *edaphology* (Brady, 1990; Tan, 1994), is concerned with the various soils and their properties as they relate to soil fertility, and plant and crop production. Analysis of soils for this purpose is called today *soil testing*, and includes plant analysis.

In the other concept of soil science, soils are studied as biochemically weathered and synthesized products in nature. This approach is known as *pedology*. Pedology includes the study of soil genesis, soil morphology, and classification. Soil chemistry serves as the connecting link between pedology and edaphology. In pedology and soil chemistry, the soil is considered an integrated mixture of air, water, and organic and inorganic components. Soil is seen not only as a source of nutrients and a medium for growing plants, but also as a necessary component for building houses, roads, industries, and disposal sites. Today, soil is also seen as a *natural resource* supporting an ecosystem

Methods of Soil Chemical Analysis

characterized by a highly diversed and interdependent population of plant and animals. Soils are in fact the product of the environment, and their properties are closely associated with environmental factors. The overuse and misuse of soils may adversely affect environmental quality.

Therefore, analysis of soils in this second concept of soil science does not entail only the analysis of the soil's nutrient status, but also includes analysis of characteristics of importance in pedology; soil organic and inorganic composition; physical, chemical and biological properties; and other attributes related to environmental quality.

4.2 SOIL TESTING

The ultimate goal of soil testing is to collect data by soil and plant analysis in order to be able to determine the nutrient status of soils and plants, which may affect the quality and yield of crops (Armstrong et al.,1984; Jones, 1985). Soil testing was designed in the beginning to detect and prevent nutrient deficiencies in soils (Baker, 1971). Over the years, it has been extended to also include soil analysis required for determining lime and fertilizers application for maintaining soil fertility levels for maximum crop yields. Currently, it is believed that soil testing is the only means for assessing nutrient deficiencies, imbalances, and/or toxicities in soils and crops, and for formulating lime and fertilizer recommendations. These purposes of today's soil testing are reflected by the four objectives reported by Fitts and Nelson (1956):

1. To group soils into classes for the purpose of suggesting fertilizer and lime practices.
2. To predict the probability of getting a profitable response to the application of plant nutrients.
3. To help evaluate soil productivity.

4. To determine specific soil conditions which may be improved by addition of soil amendments or cultural practices.

In most soil testing analysis, the sample is not weighed, but measured by volume with a standard scoop, because (1) high accuracy and precision, as frequently required in research, is not necessarily, and (2) of the time required to weigh large numbers of samples quickly. If soil samples are to be scooped rather than weighed, care must be taken in the selection of the scoop design and size. The design and size must be such that the scoop will measure out a constant volume, giving the desired weight of soil sample. For research analysis, the use of a scoop should be avoided, and the sample weighed to the desired accuracy.

4.2.1 Monitoring

Another important use of soil testing is *monitoring* (Jones, 1985), which is in fact a *preventive* soil test program. This term is used in analogy to the term *preventive care* used in human health care programs.

Monitoring is less commonly used because of the cost and large number of sampling and analyses involved. However, monitoring enables not only the detection and prevention of nutrient deficiencies, but also the maintenance of a balanced and optimum nutrient concentration in soils for obtaining high quality and maximum yields. By applying a system of periodic sampling and tracking of the analytical results, the nutrient status of soils and plants can be monitored through the growing season. Lime and fertilizer requirements can be adjusted accordingly. Specific methods of monitoring have been proposed by Jones (1983) and Jones and Budzynski (1980). Methods for monitoring special crops, such as sugarcane and peanuts, have been investigated by Clements (1960), and Jones et al.(1980). This system of monitoring necessitates consistent sampling and standardized procedures for soil and plant analysis, and is believed to be of more importance for perennial than annual

Methods of Soil Chemical Analysis 45

crops.

4.2.2 Determination of Nutrient Status

The nutrient status of soils is commonly determined by the amounts of nutrient elements extracted from soils with an extracting solution containing acids, neutral salts, or a mixture of these. A large number of extracting solutions have been suggested for this purpose, e.g., Mehlich 1 and 2 for extraction of cations in acid soils, Bray P1 for extraction of P in acid soils, and Olsen's method for extraction of P in alkaline soils. Each micronutrient element has also a specific extracting reagent, depending on pH and other soil factors (Jones, 1985). Some of the extracting agents, such as neutral NH_4OAc (1 N), Mehlich's No. 3, and the Morgan extractant, are supposed to work for extracting most of the macro- and micronutrients in a variety of soils, regardless of differences in soil properties. Such reagents are called *universal extraction reagents*. It is expected that many other reagents will be developed in the future for extraction of the nutrient elements from a wide range of soils. However, as it stands now the use of the NH_4OAc, Mehlich's No. 3, and the Morgan extractant as a *universal* extraction reagent has produced mixed results. According to Jones (1985), the availability of a universal extraction reagent would facilitate the use of multielement analyzers, such as the plasma emission spectrophotometer, and also simplify soil test analysis.

4.2.3 Plant Testing

As indicated above, plant analysis is an integral part in today's soil testing program. In analogy to the term *soil testing*, it would be appropriate to call this type of plant analysis also *plant testing*. The determination of elemental composition of cell sap is officially also called *tissue testing*, instead of *tissue analysis*. Since in most cases leaves are sampled and analyzed, plant testing can also be called *leaf testing*. Leaves are considered the most suitable part of the plants for routine elemental analysis. This type of analysis is also known as *plant tissue analysis* in other books (Jackson, 1958).

The assumption is made that the nutrient content in leaves is a reflection of the nutrient status in soils. The plant nutrient content has been found to correlate significantly with soil nutrient content, crop yield, and several crop quality factors. Therefore, the results of plant testing are used to determine the nutrient status in both soils and plants, which forms the basis for the formulation of lime and fertilizer requirements.

The nutrient content in the leaves is believed to indicate specifically the physical condition, growth, and yield of the plants. However, it was found that the elemental concentration in the leaves varies greatly with time of sampling and environmental condition. Frequently, the elements are not uniformly distributed in the leaf tissue (Sayre, 1952; Jackson, 1958; Jones, 1963). For example, B tends to be concentrated more in leaf tips, whereas Si seems to accumulate more in the leaf margins, which correlate in fact with their respective functions. Boron is needed for the growth, whereas Si is needed for providing a skeleton strength to the leaf. Jones (1970) indicated that leaves, petioles, and stems may vary considerably in their elemental makeup. Leaf blades are reported to be higher in total N, P, Na, Ca, Mg, S, Al, B, Cu, Fe, Mn, and Zn than the petioles. In addition, N, P, K, and S content in leaves usually decrease with age. This lack of homogeneity in elemental distribution necessitates sampling large amounts of leaves, or sampling the proper leaf at the proper growth stage. The selection of plant parts to be sampled should preferably be done so that (1) the nutrient status gives the best correlation with plant performance, and (2) samples are easily collected and identified. Sampling mid-shoot leaves from woody perennials and recently mature leaves from annual plants is suggested to meet the requirements above. These plant parts are believed to reflect the current nutrient status in plants, and are easy to identify and collect. The size of samples needed for analysis, on the other hand, depends on (1) the amount required for a good representation of the whole leaf and/or whole plant, (2) the concentration of the element in the leaf for good accuracy in the analysis, and (3) the sensitivity of the method of determination (Jackson, 1958).

4.2.4 Tissue Testing

Tissue testing is the determination of the elemental composition in sap squeezed from fresh plant tissue. Jones et al. (1991) indicated that it should be distinguished from *plant testing* or *plant analysis* because tissue testing involves analysis of extracted cell sap. However, many authors include in tissue testing also the determination of elements in extracts of dried ground plant materials (Ulrich, 1948; Baker and Smith, 1969; Gallaher and Jones, 1976). The latter obscures somewhat the distinction between plant testing and tissue testing.

Tissue testing is carried out mostly in the field or greenhouse by employing a rapid colorimetric analysis on cell sap extracted from plant tissue. These tests have been called *rapid field tests*, and, according to Jackson (1958), are analogous to clinical testing and diagnosis in medical science. They are usually adequate and provide a rapid and immediate answer to nutritional problems of the plants. Test kits are available for the determination of NO_3-N, P, K, and other elements. By using test papers or by adding specific reagents (provided in the kit) to the sap, a characteristic color is developed for each specific nutrient element. The elemental concentration is determined by matching the colors developed with standardized color charts. The underlying principle is that a high concentration of the nutrient element in the cell sap indicates the presence of a healthy plant. Statistical analysis between tissue testing and plant testing for NO_3-N, P, and K content in corn indicates the presence of a significant correlation (Jones, 1985).

Frequently, concurrent soil testing is carried out in conjunction with tissue testing. Various test kits for rapid soil testing are also available.

4.3 SOIL ANALYSIS IN PEDOLOGY AND SOIL CHEMISTRY

Pedology considers soil to be an ordered system of components with a construction of its own. Soil characteristics, because of their internal construction, the interaction between soil constituents, and their specific processes, can be analyzed in various ways. The soil is analyzed to indicate not only its nutrient status for crop production, but also for showing the origin, development, and transformation of soils and their constituents. Analyses are performed to study the soil's physical, chemical, and biochemical characteristics for use in crop production as well as in forestry, engineering, industry, and environmental health. Inorganic and organic soil constituents are isolated, collected, and identified, and their effect on soil conditions and the environment evaluated. The liquid and gas phase are analyzed and their application to soil-water-plant relationships examined. Currently four different approaches are generally available for these type of research analyses.

4.3.1 Traditional Soil Analysis

In the traditional approach the soil matrix is destroyed by grinding and sieving prior to the analysis. The determination of physical and chemical characteristics in this approach provides significant data on pH, water, texture, structure, soil density, organic matter content, total and exchangeable element content, soil mineralogy, and many other soil characteristics. Some of the data can be used in soil testing and soil fertility, but most of them are collected for the purpose of pedology, and soil research in general. Though several of the analytical procedures and instrumentation used here are similar to those used in soil testing, the goal or objective is different.

Methods of Soil Chemical Analysis

4.3.2 Isolation and Analysis of Soil Constituents

In this second approach the different soil components, such as obtained by particle size distribution analysis, are analyzed and their effect on soil's behavior studied. Soil and clay mineralogical analyses fall within this category. In performing the analyses, researchers use different kinds of instruments, such as the polarizing (petrographic) microscope, the ultramicroscope, and the electron microscope, as well as x-ray diffraction, differential thermal analysis, and infrared spectroscopy.

Another example of this approach is the separation of humic acids and other organic compounds. In addition to the previously stated techniques, nuclear magnetic resonance (NMR) and mass spectrometry are important tools in the analysis of humic acids.

Frequently approaches No.1 and 2 are performed concurrently.

4.3.3 Micropedological Analysis

In this third approach, the soil is studied in its undisturbed state. This method makes use of thin sections of soils, and is called *micromorphological* or *micropedological* analysis.

Undisturbed soil samples are cut with a knife, or soil clods or soil peds are collected as indicated in the sampling section of this book. Soil thin sections or polished blocks are then produced from the undisturbed samples and studied under a petrographic microscope. Different types of light are usually employed, such as ordinary light, plane polarized light, and cross polarized light. The soil components and soil properties are analyzed *in place*, and in relationship to each other. Thus, a complete understanding of the whole construction of soil and soil fabrics can be obtained.

The introduction of electron microscopy and microprobe analysis in the 1970s has created a new dimension of micropedology called *submicroscopy*. This type of instrumentation appears to be very useful, and has expanded micropedological analysis to include detailed analysis of mineral alterations and neoformations. With the help of the more modern transmission and scanning electron microscopes and with the availability of energy dispersive analysis by x-rays, also called EDAX, EDXRA, or EDA, analysis of elemental composition of a single crystal in its undisturbed state is made possible. Analysis by EDAX, which is part of x-ray fluorescence spectrophotometry, enables the determination of element concentrations in very small areas of the specimen surface, hence EDAX can also be used to identify elements and determine their concentrations in the rhizosphere, and voids or pore spaces. The fact that the minerals are identified and the concentration of the elements measured in an undisturbed state adds to the significance of the method. Such a concentration reflects one condition in the natural dynamic state of soils. Earlier this method had found extensive application in analysis of animal tissue and in medicine. With some refinement, this kind of analysis can also be adapted for soil and plant testing.

4.3.4 Analysis of Soils *in Situ*

A fourth method involves analyzing the soil in its undisturbed state in nature, in a state of motion or dynamic condition. Several of the methods in this category are well established and have been used routinely for the determination of soil physical characteristics. Determination of soil water by the neutron scattering method falls within this group. Soil samples are not collected in this technique; instead, soil moisture is measured *in situ* in the field. Other well-known analyses on soil physical properties in this category are the use of penetrometers to measure soil density, and the use of tensiometers and gypsum blocks to determine soil moisture levels directly in the field.

Currently, a new method has been introduced to analyze each soil component and its individual function and role in others *in situ* by using microscopy. This technique is still in its early stages of development and includes *micromorphometry*, a quantitative method for analysis of soil fabrics.

4.4 TYPES OF SOIL CHEMICAL ANALYSIS

4.4.1 Qualitative and Quantitative Analysis

Soil analysis comprises qualitative and quantitative analysis. The aim of qualitative analysis is the detection and identification of soil constituents or soil properties, whereas the purpose of quantitative analysis is the determination of the amounts or concentrations of soil constituents. Many of the principles and reactions of qualitative analysis can be refined and made the basis for quantitative analysis. However, a procedure proven to be effective in qualitative analysis is not necessarily useful for quantitative analysis of soils. In qualitative analysis, it is irrelevant whether all the materials contained in the sample are analyzed since the purpose is identification only. If the nature of the sample is unknown, a qualitative analysis usually precedes the quantitative analysis.

The determination of soil constituents or soil properties in both qualitative and quantitative analysis is based on the measurement of a physical or chemical property of the constituent. Therefore, three major methods can be recognized, e.g., physical, chemical, and physicochemical methods. *Physical methods* in soil analysis are methods which must involve in the analysis the measurement of a physical property of the soil constituent. However, since in soil analysis some chemistry is involved, a pure physical method, in the strict meaning of the term *physical*, is not available. Hence, it is perhaps better to distinguish the methods into *physicochemical* and *chemical* methods.

Physicochemical Methods These methods are based upon the measurement of a physical property, but a chemical reaction is an essential part of the procedure. Many physical properties can be used as the basis for the determination of a soil substance or of its concentration. An example of a physical property is the color of a solution, which is commonly used in colorimetric or spectrophotometric analysis. A chemical reaction is an essential part of the method to develop the specific color, but the final measurement is the measurement of the color, which is purely the measurement of a physical property. For instance, the colorimetric identification of phosphorus is performed by the measurement of the blue color developed by reaction with a sulfomolybdic acid-chlorostannous reducing agent. When P is absent the blue color fails to develop. The intensity of the blue color is proportional to the concentration of P in solution. Therefore, by measuring the intensity with a colorimeter, the concentration of soil P can also be determined. Another physical property frequently used is the electrochemical potential developed at the tips of electrodes. Electrometric measurement of soil pH is performed by determination of the electro-potential difference generated between the reference and indicator electrodes.

With the tremendous progress of instrumentation, numerous physicochemical methods are available today. To name a few, additional methods in this group are:

Spectroscopy In spectroscopy, excitation of soil constituents is induced by burning in a flame, arc, or spark. Emission spectra are produced by the constituents at characteristic wavelengths, which form the basis for identification. The emission intensity is proportional to the concentration, and is used in the quantitative determination of the constituent.

X-Ray Diffraction Analysis The x-ray diffraction pattern is a characteristic physical feature of crystalline compounds. It can be used for the identification of soil and clay minerals, both as a pure compound and as a component in a mixture. The relative intensity of the diffraction peaks in the xrd pattern is propor-

Methods of Soil Chemical Analysis

tional to the concentration of the mineral, thus making quantitative determination possible.

Chemical Methods   Chemical methods depend on the application of a chemical reaction involving the soil constituent being analyzed. Three major types of methods can be distinguished in this category: (1) gravimetric method, (2) volumetric method, and (3) gasometric method.

Gravimetric Method   By this method the soil constituent or its reaction product is found by weighing. It is an important method in (1) soil moisture determination, (2) particle size distribution analysis, (3) bulk density measurement, (4) particle density measurement, and (5) dry weight determination of plants.

Volumetric Method   By this method the amount of soil constituent is determined by the amount of reagent required to react with it. Based on the type of chemical reaction, volumetric methods are distinguished into: (1) acid-base titrations, (2) methods based on precipitation and complex reactions, and (3) methods based on oxidation-reduction reactions.

In *acid-base titrations* a solution of the substance being analyzed is treated with a solution of a chemical reagent of exactly known concentration, called *standard reagent*. This standard reagent is added from a buret to the solution containing the sample until the amount added is equivalent in amount to the substance analyzed. This point is called the *theoretical endpoint* or the *equivalent point*. An indicator is added to the system to detect the *endpoint of titration*. The latter does not necessarily to be equal to the equivalent point, and the difference is referred to as the *titration error*. Preferably a suitable indicator must be selected so that the titration error is negligibly small.

In acid-base titrations involving reactions with monovalent

ions, 1 equivalent equals 1 mole of substance. However, in reactions with polyvalent ions, the equivalent weight of the substance is found by the reaction process:

$$H_3PO_4 \rightarrow H^+ + H_2PO_4^- \qquad \text{1 equivalent = 1 mol}$$

$$H_3PO_4 \rightarrow 2H^+ + HPO_4^{2-} \qquad \text{1 equivalent = 1/2 mol}$$

$$H_3PO_4 \rightarrow 3H^+ + PO_4^{3-} \qquad \text{1 equivalent = 1/3 mol}$$

In *complex reactions*, the standard reagent is added until the substance being analyzed is completely complexed. For example in the determination of cyanide, cyanide is titrated with silver until all the cyanide is complexed by the silver. If titration is performed according to Mohr's method, the reaction occurs as follows:

$$Ag^+ + CN^- \rightarrow AgCN$$

Here the equivalent equals 1 mol. If Liebig's method is used in the titration, the end point of titration is reached when the following reaction has taken place:

$$Ag^+ + 2CN^- \rightarrow Ag(CN)_2^-$$

In this case, two CN^- are complexed, hence 1 equivalent = 2 mol.

In *oxidation-reduction* reactions, the equivalence of a substance is found from the change in oxidation state or from the number of electrons transferred during the reaction:

1. In the titration of ferrous into ferric iron using an oxidation agent, the state of oxidation of iron changes from 2 to 3:

$$Fe^{2+} \rightarrow Fe^{3+} + e^-$$

Therefore, the equivalent of ferrous iron equals 1 mol. Note that only 1 electron is released during the reaction.

2. In volumetric analysis, when $KMnO_4$ is used as an oxidizing agent in acid medium, the permanganate ion is reduced into manganous ions:

$$MnO_4^- \rightarrow Mn^{2+} \quad \text{or} \quad Mn^{7+} \rightarrow Mn^{2+} + 5e^-$$

The change in oxidation state is from 7 to 2 (or 5 electrons have been transferred during the reaction), which means a change in five units. Hence, the equivalent weight of permanganate is 1/5 mol. This type of titration finds application in the determination of soil organic carbon. The Walkley-Black method uses $K_2Cr_2O_7$ and H_2SO_4 in the determination of organic carbon. The dichromate ion is reduced to the chromic ion according to the reaction:

$$Cr_2O_7^{2-} + 14H^+ + 6e^- \rightarrow 2Cr^{3+} + 7H_2O$$

The oxidation state of dichromate changes with 6 units. Correspondingly 6 electrons have been transferred during the reaction. Therefore, 1 equivalent of Cr_2O_7 equals 1/6 mol.

CHAPTER 5

DETERMINATION OF SOIL WATER

5.1 DEFINITIONS AND PRINCIPLES

Water, a renewable resource, belongs to a gigantic hydrological cycle. It continuously moves at the earth's surface and evaporates from the earth into the atmosphere. After reaching the atmosphere, water is moved by the wind until it finally falls to the ground again as some kind of precipitation. In the cycle, part of this water on the ground is evaporated directly from the soil or runs off the surface, called runoff, into streams, lakes, swamps, and oceans. Another part is adsorbed by the soil or percolated through the soil to reach the ground water and springs.

Water is held in the soil by both adhesive and cohesive forces. *Adhesion* is defined as the attraction of the solid soil particles for water, whereas *cohesion* relates to the mutual attraction between water molecules. Another force affecting retention and movement of water in soil is the *capillary* force, by which water is adsorbed in the micropores or capillaries. The capillary force depends on the surface tension of water. The water retained in the soil by these forces constitutes the water content in soils, and includes the reserve water supply for plant

Determination of Soil Water

use between additions of water by rain or irrigation. The resulting force that holds water in soils is called *soil moisture tension*. The tensional force may be expressed in a variety of ways, e.g., kg per m^2, centimeters of water or mercury, atmospheres or bars, and pF. The basic unit, however, is the *cm of water*, which is defined as the force equal to the weight of the height of a water column in cm. It is interpreted in the same way that the cm Hg for barometric pressures is interpreted. The values for soil moisture tension in cm of water range from 0 to 10,000,000 cm. A moisture tension of 0 cm water indicates that there is no tension, which is attributed to the presence of excessive amounts of water in soil. A soil moisture tension of 10,000,000 cm means that water is held with a very large force, in other words there is not much water in the soil, or the soil is dry. The unit cm H$_2$O can be converted into bars by dividing by 1,000. Hence,

$$0 \text{ cm } H_2O = 0/1{,}000 = 0 \text{ bar}$$
$$10{,}000{,}000 \text{ cm } H_2O = 10{,}000{,}000/1{,}000 = 10{,}000 \text{ bars}$$

The pF unit is also derived from the soil moisture tension in cm H$_2$O by taking the logarithm as follows:

$$pF = \log (\text{cm } H_2O)$$

The minimum value for cm H$_2$O to be used in the log equation is then a soil moisture tension of 1: pF = log 1 = 0. The maximum value is pF = log 10,000,000 = 7.

The pF is unique to soil moisture expression, although atmospheres or bars are also used. Water can exist in soil under a tension varying from pF = 0 (no tension) to pF = 7.0 (high

Table 5.1 Soil Moisture Constant and their Corresponding Tension Values

Soil moisture constant	cm H$_2$O	Bars	pF	MPa
Maximum retentive capacity	0	0	0	0
Field capacity	346	0.3	2.54	0.03
Moisture equivalen	1,000	1	3.0	0.1
Wilting point	15,849	15	4.2	1.5
Hygroscopic coefficient	31,623	31	4.5	3.1
Air dry soil	1 x 10^6	1,000	6.0	100
Oven dry soil	1 x 10^7	10,000	7.0	1,000

tension). In the presence of excess water, there is practically no tension and the pF = 0 (Table 5.1). The point at which water is held in soils after excess water has drained by gravity is called *field capacity* (pF = 2.54). At field capacity, the soil contains maximum amounts of water readily available to plants. Plants use little gravitational water since this water rapidly drains away to a depth below the root zone. As the soil dries and the water content decreases, the degree of soil moisture tension becomes increasingly larger. Therefore, the force that the plant has to exert to remove the remaining water increases as the soil dries. Eventually a point is reached at which the force exerted by the plant is not sufficient to extract water at a sufficient rate for growth. At this point, the so-called *wilting point* (pF = 4.2), the plants start to wilt. The amount of water available to plants, called simply *available water*, is the amount of water held in soil between the field capacity and the wilting point. In view of the foregoing, three general types of soil water measurements will be discussed here: measurement of total water content, measurement of available water content, and

Determination of Soil Water

measurement of soil moisture tension.

5.2 DETERMINATION OF TOTAL SOIL WATER CONTENT

The scientist must ascertain the total water content of soils in order to conduct research in edaphology, pedology, soil chemistry and engineering, and other branches of soil science. It can be determined directly and indirectly.

5.2.1 Direct Determination of Water Content

The simplest method for the direct determination of soil water content is the gravimetric method, which in principle involves the measurement of water lost by weighing a soil sample before and after it is dried at 105-110°C in the oven. The result is presented in the percentage of water, which can be expressed in a dry mass percentage, a wet mass percentage, or a volume percentage. Each of these weight percentages can be calculated using the equations given below:

1. Dry mass % H_2O = (mass H_2O/mass oven-dry soil) x 100 %

2. Wet mass % H_2O = (mass H_2O/mass moist soil) x 100 %

3. Volume % H_2O = dry mass % H_2O x bulk density

The dry mass percentage is used in soil science and chemistry. The wet mass percentage is used in the botanical sciences and finds practical application in agriculture, horticulture, and nurseries in the weighing of fresh produce and vegetables.

Procedure

Place the soil sample in a clean preweighed stoppered weighing flask or bottle. The amount used should be approximately 5 - 10 g, and it should be spooned quickly in the flask. Close the flask with the ground stoppered lid, and carefully weigh it to the nearest 1 or 0.1 mg, depending upon the accuracy desired. Remove the lid from the flask, and dry the flask with its content at 105 -110°C for 24 h (or overnight) in a forced draft oven. At the end of the 24-h period (or the next morning), allow the flask to cool in a desiccator. Place the lid back on the flask and carefully weigh the flask with its content to the nearest 1 or 0.1 mg. The amount of water lost, which is the water content of the sample, may be calculated as follows:

$$H_2O \text{ lost} = \text{weight of moist soil} - \text{weight of ovendry soil}$$

Example of Calculations

Weight of flask + lid + soil before drying = 12.1234 g
Weight of flask + lid + soil after drying = 10.4321 g

Amount of H_2O lost = 1.6913 g

Dry mass percentage: $\%H_2O = 1.6913/10.4321 \times 100\% = 16.2\%$

Wet mass percentage: $\%H_2O = 1.6913/12.1234 \times 100\% = 13.9\%$

If bulk density value is determined and equals 1.5 g/cc, the volume percentage is:

$$\%H_2O_v = 16.21 \times 1.5 = 24.3\%$$

5.2.2 Indirect Determination of Water Content

The indirect determination of soil water is often performed by using the gypsum block method, or porous block method. A gypsum block is a block of $CaSO_4$ or casting plaster (Bouyuocos, 1953), in which two electrodes are imbedded. A porous block has a similar design but is composed of a variety of materials, e.g. plastic and fiberglass (Coleman and Hendrix, 1949; Gardner, 1986). With both types of blocks, the measurements are performed directly in undisturbed soil in the field, a procedure which offers the scientist a distinct advantage. The water content is measured through the electrical resistance of the block. The rate of electrical resistance, measured by a voltmeter, is proportional to the amount of H_2O adsorbed by the block. Since the amount of adsorbed H_2O is a function of the amount of water in the soil, the rate of electrical resistance can be used to determine the amount of water in the soil.

Procedure

Prior to measurement of water in the field, the block has to be calibrated, which is done as follows. Soak the block with water and separately moisten a soil sample to a state that it can be packed firmly around the block. After packing the block with the moistened soil, determine the resistance of the block. At the same time, take a sample of the moistened soil and measure its water content by the gravimetric method as discussed above. Calculate its dry mass percentage.

Repeat this calibration analysis several times at different levels of water content. Plot the values for resistance against the water contents to obtain a calibration curve.

The calibration analysis can also be conducted under the influence of suction forces by using a suction or vacuum pump. This technique generates a calibration curve which includes points for field capacity and wilting point, respectively. Such a curve is expected to be very useful in the scheduling of irrigation times on farmland and turf greens.

To measure the water content of the soil in the field, a hole must be dug vertically from the surface soil downwards. The block can be placed at any depth desired (Figure 5.1) or it can be placed horizontally in the side of a soil horizon of an exposed soil profile. Wet the block thoroughly before placing it in the hole, after which it is packed thoroughly with soil for good contact. Allow the block to equilibrate with the soil overnight, and measure the resistance. Find the water content from the calibration curve.

5.3 DETERMINATION OF AVAILABLE WATER CONTENT

As defined above, the available water content is the amount of water held by soils between the field capacity and wilting point. At field capacity, the soil has a pF = 2.45, which corresponds to a pressure or suction force of 0.3 bars. On the other hand, at wilting point, the soil exhibits a pF = 4.2, which corresponds to a suction force of 15 bars. By applying these suction forces to moist soil, the soil moisture contents at both field capacity and wilting point can be measured. Frequently, it is sufficient to measure only the soil moisture content at field capacity to get an indication of available water content. This will simplify the analysis.

5.3.1 Pressure-Plate Method

Procedure

This method requires the use of a pressure-plate or tension plate instrument (Figure 5.2). Place a cellulose membrane on the porous porcelain plate in the instrument. Fill the retainer rings in the instrument with a known (preweighed) amount of

Determination of Soil Water

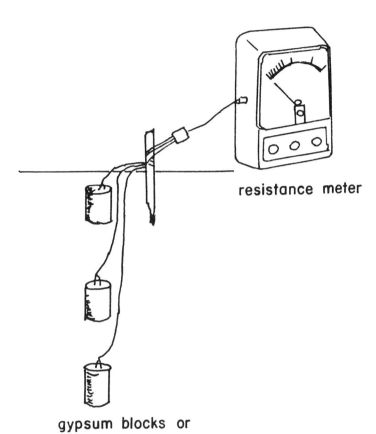

Figure 5.1 Gypsum blocks placed at different depth in the soil profile.

Figure 5.2 Pressure-plate apparatus for determination of soil water at 0.3 to 15 bars pressure.

Determination of Soil Water

soil sample, and fill the plate with water to wet the samples from below. Cover with a plastic sheet and allow the sample to soak overnight. After the soil sample is saturated with water, remove the excess water. Close the instrument and tighten it with the screws. For field capacity water, apply a pressure of 0.3 bars until no more water is forced out of the sample, which usually takes several days. As soon as no more water can be forced out, quickly transfer the moist sample to a container. Record the weight of the container with and without moist soil. Dry the moist soil in the container in a forced-draft oven at 105°C for 24 h, and calculate the moisture content as percentage of oven-dry weight soil.

The analysis can be repeated at a suction force of 15 bars for the determination of water at the wilting point. The available water content is then the difference between the water at field capacity and the wilting point.

5.3.2 Alternative Methods

In the absence of a pressure-plate instrument, either of the following two methods can be used with regular glassware. The first alternative is a suction method that requires the use of a vacuum pump. For safety reasons, the determination of water at 15 bars of pressure is not done with this method, but is performed separately with the pressure-plate method as discussed above. The second alternative method is simpler and does not use a vacuum pump. However, it measures only water at field capacity.

Alternative Suction Method
Procedure

Place in a preweighed 100 mL beaker containing a stirrer (= A g) approximately 25 g of soil sample. Weigh the beaker, stirrer, and soil accurately to the nearest 0.1 mg (= B g). Prepare a saturated paste of the soil above by adding distilled water, and mix with the stirring rod. The first addition of water should be about 5 mL, which is followed by smaller amounts of water until a saturated soil paste is attained. Mix the water thoroughly after each addition. At saturation, the soil glistens as it reflects light, and flows slightly when the beaker is tipped. The paste will slide cleanly and freely off the stirring rod. Reweigh beaker, stirring rod, and saturated soil (= C g).

Transfer the saturated soil paste as much as possible to a Buchner filter containing a moistened filter paper. Connect the Buchner filter to a preweighed vacuum Erlenmeyer flask (= D g). Reweigh the beaker, stirring rod, and soil residue promptly after the transfer (= E g). This reweighing is needed to determine the amount of soil transferred into the Buchner filter.

Apply a suction with the vacuum pump at 0.3 bars suction or 25 cm Hg. Continue extracting water by suction until all the water that will yield to this suction force has been removed, or when the dripping ceases. If the soil texture is sufficiently coarse, most of the removable water will have been drawn from the sample within 30 minutes. Disconnect the Erlenmeyer flask and weigh it with its content (= F g). The remaining water that has withstood a suction force of 0.3 bars is considered water at field capacity.

If desired, the procedure above can be repeated with a suction force of 15 bars for the determination of the wilting point. Care and caution must be exercised so that the glassware, which is subjected to a pressure of 15 bars, does not break. This is one reason why only field capacity measurements are done with this alternative method. The safest method for determining the wilting point is the pressure-plate method; with this method, an accident is very unlikely.

Method of Calculations

Amount of air-dry soil = B - A = X g

Determination of Soil Water

Weight of the soil above on an oven-dry basis:
$(X \times 100)/(100+\%H_2O)$

Water content in soil sample = X - wt oven-dry soil = Y g

Amount of water at saturation: $(C - B) + Y = Z$ g

% H_2O at saturation = (Z/wt oven-dry soil) x 100% = P%

Amount of saturated soil transferred = $C - E = S$ g

Amount of this soil on oven-dry basis =
$(S \times 100)/(100 + P\%) = Q$ g

Amount of water extracted at 0.3 bars = $F - D = R$ g

Amount of water that withstood 0.3 bars suction:
$(S - Q) - R = T$ g

% H_2O at field capacity (0.3 bars suction) = T/Q x 100%

The amount of available H_2O = %H_2O at field capacity - %H_2O at wilting point.

Measurement of Field Capacity Water
Procedure

Fill a preweighed 100 mL graduated cylinder (= A g) with soil, and pack the soil by gently tapping the cylinder with your hand 10 times. Continue filling the cylinder and tapping it until a tapped volume of 100 mL soil is obtained. Weigh the cylinder and 100 mL soil (= B g). Measure 10 mL of water and add this slowly to the soil in the cylinder. Observe the nature of the wetting front. After percolation and capillary adjustment have taken place for 24 hours, measure the volume of soil wetted by the water (= C mL).

Method of Calculation

Amount of air-dry soil = B - A = X g

Amount of soil on an oven-dry basis = (X x 100)/(100+%H_2O) = Y g

Dry mass %H_2O has to be determined separately (= Z g).

Bulk density of soil = Y/soil volume = Y/100 = K g/cc

Volume of soil moistened by 10 mL H_2O = L mL

Oven-dry weight of the amount of soil above = L x K = M g

Amount of water at field capacity = (M x Z) + 10 = P g (assume 10 mL water = 10 g)

Percentage of water at field capacity = (P/M) x 100%

5.4 DETERMINATION OF SOIL MOISTURE TENSION

As defined earlier, the tension or suction is a resultant force with which water is held in soils. This tension is related to the water potential. Therefore, by measuring soil moisture tension, the force holding water is measured, and the value obtained may show indirectly the relative amount of water present in the soil. Soil moisture tension can be measured in a variety of ways. It can be measured by the pressure-plate method described above, but a method often used in the field is the tensiometer

Determination of Soil Water

method. A tensiometer is composed of a porous porcelain cup attached to a plastic tube. It is filled with water and then placed in the soil. The tension is traditionally measured by a gauge connected to the stem of the tensiometer. However, commercial tensiometers are currently available equipped with a portable measuring device called a *tensimeter* replacing the gauge (Figure 5.3). The tensimeter yields automatic readouts in mbars. Converting mbars into pF values proceeds as follows:

1,000 cm H_2O pressure = 1 bar
1 cm H_2O pressure = 1 mbar
pF = log cm H_2O pressure

A list of pF values of importance in soil science and agriculture, and their equivalents in bars and cm H_2O, are provided in Table 5.1.

Procedure

Prior to using it in field measurements, the tensiometer may be tested first by placing it upright in a pail of water. Remove the cap and allow water to accumulate in the barrel of the porous porcelain cup. If no water is moving through the walls of the cup, the cup is clogged and must be cleaned. After it has passed the test above, fill the tensiometer with de-aerated water. De-aerated water is produced by boiling and cooling the water. Create a vacuum of 0.6 to 0.8 bars with a vacuum pump to remove air bubbles trapped in the tensiometer (Cassell and Klute, 1986).

After this testing procedure, the tensiometer is ready for use in the field. Dig a hole with an auger, and pour water in the hole to make a soil slurry at the bottom of the hole. Insert an

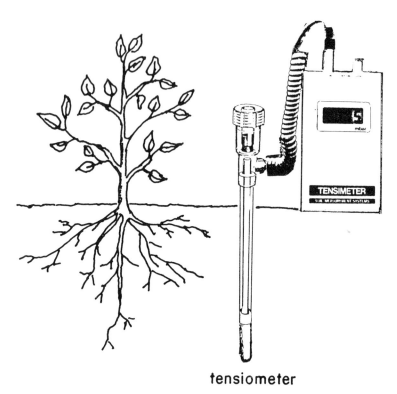

tensiometer

Figure 5.3 Tensiometer for measurement of soil-moisture tension with a tensimeter.

Determination of Soil Water

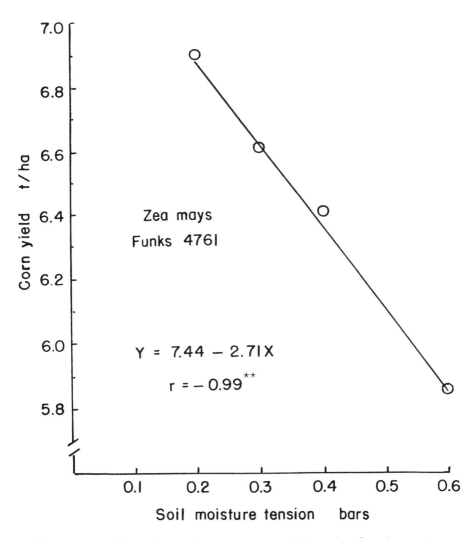

Figure 5.4 The relation between corn yield and soil moisture tension (from unpublished report of the University of Georgia, Agricultural Experiment Stations).

empty tensiometer and seat it by twisting it into the soil slurry. Pack the hole around the cup and barrel with soil for good contact, and fill the tensiometer with de-aerated water. Allow it to come to equilibrium with the soil environment for 6 hours. Plug in the tensimeter to the tensiometer and take a reading. A series of readings can be made over periods of time with this method. The mbars may be used as is to indicate the soil moisture tensions, or they can be converted into pF values if desired.

An illustration of the use of tensiometer analysis in agriculture is given in Figure 5.4. Corn plants were grown in soils at different water tensions, and corn yield decreased as moisture tension increased. As indicated above, as the water content decreases, the value for soil moisture tension increases. Hence water stress increases since plants have to exert an increasingly greater force to extract water from the soil.

CHAPTER 6

DETERMINATION OF SOIL TEXTURE

6.1 DEFINITION AND PRINCIPLES

Physically, the soil is a mixture of mineral matter, organic matter, water and air. The mineral matter is composed of inorganic particles varying in size from stone and gravel to powder. These inorganic particles, separated according to size, are referred to as *soil separates*. The United States Department of Agriculture (Soil Survey Staff, 1951) recognizes three major groups of soil separates: sand (2.0-0.050 mm), silt (0.050-0.002 mm), and clay (<0.002 mm). The three fractions can be subdivided into finer size fractions as follows (Jackson, 1956):

	Diam.
Very coarse sand:	2.00 - 1.00 mm
Coarse sand:	1.00 - 0.50 mm
Medium sand:	0.50 - 0.25 mm
Fine sand:	0.25 - 0.10 mm
Very fine sand:	0.10 - 0.05 mm
Coarse silt:	0.050 - 0.020 mm
Medium silt:	0.020 - 0.005 mm
Fine silt:	0.005 - 0.002 mm

Coarse clay: 2.0 - 0.2 µm
Medium clay: 0.2 - 0.08 µm
Fine clay: < 0.08 µm

The relative proportions of the soil separates determine the soil texture. When clay is present in dominant proportions relative to silt and sand concentrations, the soil is described as having a fine or heavy texture. Fine textured soils are plastic and sticky when wet, and hard and massive when dry. They are very difficult (heavy) to plow, hence the term *heavy texture*. On the other hand, when sand is dominant, the soil exhibits a coarse or light texture. Coarse textured soils are loose and friable and are easy to plow, hence the term *light in texture*. In the presence of balanced concentrations of sand, silt, and clay, the soil is considered to have a medium texture.

Texture is an important characteristic of soil, affecting drainage conditions, water-holding capacity, amount and size of pores, and plant root development. Consequently, the rate of water intake, water supplying power, aeration, and soil fertility are all influenced by soil texture. The USDA (Soil Survey Staff, 1951) recognizes twelve types of soil textures (Figure 6.1) for classifying soils into twelve soil classes. Because of this, the terms soil textures and soil classes are frequently used interchangeably. However, soil texture is a soil characteristic, whereas soil classes are groups of soils differentiated by differences in soil texture.

Two groups of methods are present for determination of soil texture, e.g., (1) field or rapid method, and (2) laboratory method. By the field method, a moist sample and wet soil sample are squeezed between the fingers. The presence of clay is indicated when the soil feels sticky when wet, and, by the way, the moist sample *slicks out* to form a thin ribbon. Sand is present when the sample feels gritty, whereas the presence of silt produces a slick, soapy feeling. The accuracy of this type of

Determination of Soil Texture 75

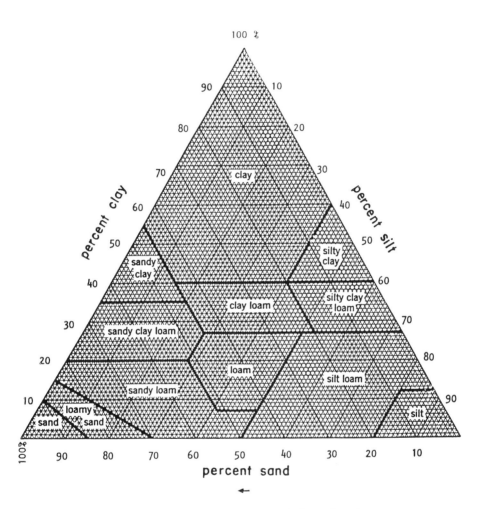

Figure 6.1 USDA guide for soil textural classification.

soil texture determination depends on skill and experience.

The laboratory method of soil texture determination, an indirect method, is conducted through the quantitative determination of soil separates. Today, this type of analysis is called *particle size distribution* or *particle size analysis* (Soil Survey Staff, 1972; Gee and Bauder, 1986). Formerly called *mechanical analysis*, this procedure sorts out the inorganic soil particles into the sand, silt, and clay fractions. Two important steps in the analysis are dispersion and sedimentation.

Dispersion A complete dispersion of the soil sample is one of the requirements of this analysis. In most soil samples, the sand, silt, and clay particles are cemented together into aggregates or granules by organic matter, clay and salts. These aggregates must be disrupted into the individual particles and the latter suspended in water, a process called *dispersion*. The dispersion must be strong, but not so drastic as to crush the individual particles. Usually a chemical reagent, such as NaOH or $NaPO_3$ (sodium metaphosphate or calgon), is added to enhance dispersion. Final dispersion is effected by thorough shaking or mixing mechanically with a blender.

Sedimentation After the soil has been dispersed, the different particle size fractions are sorted out by sedimentation techniques using the *Bouyoucos hydrometer* or the *pipette method*. In the Bouyoucos hydrometer method (Bouyoucos, 1962), the amount of the different size fractions is determined with a *soil hydrometer* (Figure 6.2), which is designed to measure the amounts in grams of particles in suspension. In the pipette method, a pipette is used to withdraw an aliquot of the soil suspension at a predetermined time and depth, then dried and weighed. For quantitative collection of the various size fractions, the suspended material may be siphoned off at fixed time periods and dried. All these methods of measuring and collection of soil separates are based on the differential rate of

Determination of Soil Texture 77

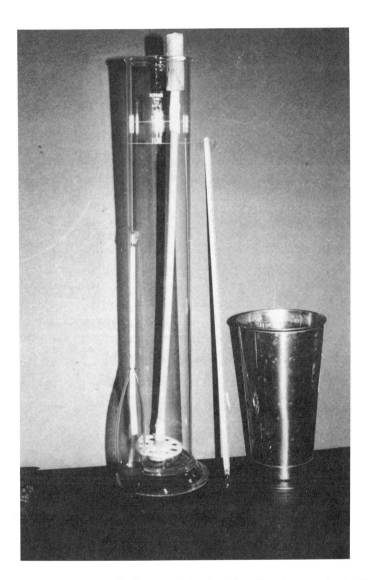

Figure 6.2 Tools for particle size distribution analysis: Hydrometer, A.S.T.M. cylinder, stirrer, thermometer, and steel beaker or blender cup.

settling of particles during sedimentation, as formulated by the *law of Stokes*:

$$V = [2 r^2 (d_p - d_w) g]/9\eta$$

in which V = rate of settling (cm/s) of particles with an effective radius = r (cm) and density = d_p (g/cm^3) falling through a liquid medium with density = d_w (g/cm^3) and viscosity = η (poise, g/cm/s) under the acceleration of gravity = g (981/sec^2).

Stokes' law indicates that V is a function of the radius, hence the larger the particles the faster the rate of settling. Therefore, sand particles will settle out first, followed by silt particles. Clay particles tend to remain in suspension for a longer time.

Since d_p, d_w (d of water = 1 g), and η are constants for a given temperature, $[2(d_p - d_w) g]/9\eta$ = k (constant). Consequently, the law of Stokes can be reduced into:

$$V = kr^2$$

The value k is dependent on the temperature. When water is used as the dispersion medium, the value k = 6000 at 25°C. Using the modified version of the law of Stokes, Bouyoucos (1962) found that all particles >0.050 mm (sand) settle out after 40 s, whereas only clay remains in suspension after 2 h.

6.2 HYDROMETER METHOD.

The hydrometer method is a simple and rapid method for the analysis of soil texture through the measurement of total sand (2.0-0.05 mm), silt (0.050-0.002 mm), and clay (<0.002 mm)

Determination of Soil Texture

contents.

6.2.1 Dispersion Reagent

Sodium metaphosphate, $(NaPO_3)_x \cdot Na_2O$, ($x \sim 13$).

Dissolve 40 g of the reagent in 1 L of distilled water. Use 10 mL of this solution in the analysis.

6.2.2 Procedure

Soil moisture content must be determined separately on the soil analyzed, since all results must be calculated on an oven-dry basis of soil.

Weigh 50.0 or 100.0 g of soil sample and transfer it quantitatively into a blender cup. For proper hydrometer readings, 50 g is used if the soil is clayey, whereas 100 g is used if the soil is sandy in nature.

After the soil sample has been transferred, fill the blender cup with distilled water to within 10 cm of the top (rim) and add 10 mL of sodium metaphosphate solution.

Instead of sodium metaphosphate, a 2-4 M NaOH solution can also be used, which should be added dropwise, under constant stirring, until the soil suspension has a pH=10-11.

Attach the cup to a blending or stirring machine and blend mechanically for 15 minutes. Transfer the soil suspension into an A.S.T.M. soil testing cylinder. Wash the remaining soil residue quantitatively in the cylinder by spraying with water from a water bottle. If 50 g of sample is used, make up the volume in the cylinder with water to the 1130 mL level. If a 100 g sample is used, fill the cylinder to the 1205 mL level. Mix the suspension thoroughly by stirring with a stirring rod so that all sediment disappears from the bottom of the cylinder. Record (*clock*) the exact time when stirring is stopped.

Carefully place a hydrometer into the suspension, and exactly 40 s after the stirring is stopped, read to the nearest 0.5

scale division the top of the meniscus on the hydrometer stem. Remove and rinse the hydrometer. Stir the suspension again and repeat the analysis of the 40 s reading. The average of the two readings is taken as the result, which equals the amount of silt + clay in grams. Determine and record the temperature of the suspension after you have removed the hydrometer.

Stir the suspension again thoroughly. Take a third hydrometer and temperature reading after 120 minutes (2 h) of settling time. This reading will measure the amount of clay in grams. Clean and dry all equipment.

Since most hydrometers in the U.S. are calibrated at 68°F (seldom in °C), all hydrometer readings must be corrected if the temperature of the suspension is higher or lower than 68°F. After every hydrometer reading, the temperature of the suspension must be taken, since for each 1°F variation from 68°F a temperature correction of 0.2 graduation must be applied. For temperatures >68°F, the correction should be added to, and for temperatures <68°F the correction should be subtracted from the actual hydrometer reading. An example of such a calculation is given below.

Example of Calculation

52.5 g soil (% H_2O = 5.0%) were used for particle size distribution analysis. The calibration temperature of the hydrometer is 68°F.

The amount of soil is figured on an oven-dry basis:

(52.5 x 100)/(100 + 5) = 50g.

Determination of Soil Texture

	Hydrometer reading	Temperature	Corrected reading
40 s	25.0	78°F	25.0 + 0.2(78 - 68) = 27.0 g
2 h	10.0	76°F	10.0 + 0.2(76 - 68) = 11.6 g

The basic principle in the following calculations is:

% sand + % silt + % clay = 100%

40 s reading: %(silt + clay) = (27.0/50) x 100 = 54.0%
%nbsp;sand = 100 - 54.0 = 46.0%

2 h reading: % clay = (11.6/50) x 100 = 23.2%
% silt = 54.0 - 23.2 = 30.8%

Soil class name (from USDA textural triangle): Loam

6.3 CENTRIFUGE METHOD

The centrifuge method is an elaborate but rapid method for the determination of soil texture. With this method, the various sub-fractions of sand, silt and clay can be determined. In this method organic matter and other cementing agents are removed from the soil sample prior to the analysis.

6.3.1 Reagents

Hydrogen peroxide, H_2O_2, 30 to 35%, is used to remove organic matter.

Sodium hexametaphosphate, $(NaPO_3)_x \cdot Na_2O$, is used to disperse the soil sample (see the hydrometer method).

6.3.2 Procedure

Removal of Organic Matter Weigh 10.0 g (or more if desired) of soil sample in a 500-mL beaker. Wet the soil with distilled water. Place the beaker on a hot plate, and add 25 mL of H_2O_2 and 10 mL of distilled water. Cover with a watchglass, and allow the H_2O_2 to react with the sample. After the frothing and bubbling have ceased, add an additional 25 mL H_2O_2 and allow the reaction to run again to completion. After the organic matter is completely oxidized, remove the remaining H_2O_2 from the mixture by heating very gently for 15 minutes. Do not heat until dryness.

Removal of Dissolved Material and Dispersion Wash the soil several times with water to remove the dissolved mineral matter produced by the oxidation of organic matter. Fill the beaker half full with distilled water, and allow the soil particles to settle. Siphon and discard the clear (often colored) supernatant carefully without disturbing the settled soil sample at the bottom. Repeat this washing procedure twice. Then transfer the sample quantitatively into a blender cup. Fill the blender cup with water to approximately 10 cm below the rim. Add 10 mL of the dispersion reagent, and blend the soil mixture for 15 minutes.

Separation of Sand Insert a funnel into a 500-mL beaker and place a 0.050 mm sieve over the funnel. The soil mixture in the blender cup is then poured quantitatively into the sieve. The remainder at the bottom of the beaker is sprayed with water onto the sieve. Wash the sample on the sieve by gently jet spraying it with distilled water from a water bottle. The mate-

Determination of Soil Texture

rial passing through the sieve, and collecting in the beaker, contains the silt and clay fractions. The total sand fraction, remaining on the sieve, is dried, and weighed. The weighed dry sand is then transferred into the top sieve of a stack of sieves, composed of 1.0 mm, 0.50 mm, 0.25 mm, and 0.10 mm sieves, respectively, from top to bottom. Shake the stack of sieves mechanically with an oscillating shaker for 30 minutes. Record the weights of the individual sand fractions. If the sum of the weights of the different size fractions equals the weight of the total sand fraction, the analysis is correct. Convert the results in percentages on the basis of oven dry weight of soil.

Separation of Silt The suspension in the beaker, containing the silt and clay, is dispersed again, stirred, and allowed to stand for 5-10 minutes (Jackson, 1956) in order to separate the coarse silt fraction. The supernatant suspension is carefully decanted and transferred into a 500-mL beaker. The coarse silt fraction (0.050-0.020 mm) sedimented at the bottom is collected, dried and weighed. To separate the remaining silt from the clay fraction, the supernatant is centrifuged for 3 minutes at 300 rpm with the *International* centrifuge. The supernatant containing the clay fraction is carefully decanted for further analysis into coarse, medium and fine clay. The residue sedimented by centrifugation, composed of the medium and fine silt fractions, is dispersed again. This step is followed by centrifugation for 3 minutes at 300 rpm, which results in the separation of these silt fractions. The supernatant (fine silt) and the sediment (medium silt) are collected, dried and weighed. The total silt content, needed for soil texture determination, can be calculated again on an oven-dry basis of soil.

Separation of Clay In order to measure the clay content, the remaining suspension can be dried and weighed. This process determines the total clay content, which should be expressed in terms of percentages on an oven-dry basis of soil. Frequently, it is necessary to study the finer subfractions of clay. To obtain the various subdivisions of clay, the clay suspension remaining after the separation of silt is centrifuged with the International centrifuge for 30 minutes at 2400 rpm (Jackson, 1956). The coarse clay fraction (2.0-0.2 µm), at the bottom of the centrifuge flask, is collected, dried and weighed. The supernatant is centrifuged with a superspeed centrifuge for 15 minutes at 50,000 rpm. Both the supernatant (<0.08 µm fraction or fine clay) and the sediment (0.2 - 0.08 µm fraction or medium clay) are collected, dried and weighed.

Determination of Soil Texture The percentages of total sand, total silt and total clay are projected in the USDA textural triangle to determine the soil class.

6.4 PIPETTE METHOD

The pipette method is an alternative method for the determination of soil texture. It is, however, time consuming, and the accuracy of the results approximates that of the hydrometer and centrifuge methods.

6.4.1 Procedure

The procedure greatly resembles that of the centrifuge method. The sample is processed similarly by dispersion, sedimentation and decantation, as in the separation of sand and coarse silt. The remaining suspension, containing medium and fine silt and clay, is then analyzed at a constant temperature by taking samples with a pipette. The suspension is first transferred to a 1-L cylinder, which is placed in a waterbath at room temperature (25°C). After dispersion, the volume is made up to 1 L with distilled water. The suspension is stirred thoroughly. After sedimentation is allowed to occur for 4.3 minutes (Kilmer and Alexander, 1949; Wirjodihardjo and Tan, 1964; Gee and Bauder, 1986), an aliquot is drawn with a

Determination of Soil Texture

50-mL pipet at a depth of 10 cm. The content, containing the medium silt fraction (20 - 5 µm), is dispensed into a preweighed porcelain crucible, dried, and weighed. A second aliquot is drawn after 68.3 minutes of sedimentation time, dried, and weighed. This process determines the fine silt fraction (5 - 2 µm). The clay fraction is obtained by pipetting an aliquot at a depth of 7 cm after a sedimentation time of 5 h.

Calculations of the Pipetted Fractions

$$\% \text{ Pipetted fraction} = \frac{(W - D)}{S} \times \frac{(1000)}{50} \times 100\%$$

where W = weight of pipetted fraction, D = weight of dispersing agent in 50 mL, and S = weight of soil on an oven-dry basis.

CHAPTER 7

DETERMINATION OF SOIL DENSITY AND POROSITY

7.1 DEFINITION AND PRINCIPLES

The density of soil refers to the mass of soil per unit volume of soil (Spangler and Handy, 1984; Brady, 1990). It is a means of expressing the amount of soil, and its value depends partly on the mineral and organic matter content, and partly on the amount of pore spaces or soil porosity. Soil density is an important physical property that affects both agricultural and engineering operations. The denser the soil, the less permeable the soil. A dense, compacted soil inhibits plant growth, yet it provides the engineer and the architect with a strong support for building foundations. Soil mass exerts pressures on dams and tunnels, and contributes to landslides. The mass of a given volume of soil can be expressed in terms of bulk and particle density.

Bulk density is defined as the mass (weight) of soil per unit volume of undisturbed soil or bulk soil volume. Thus:

Bulk density = weight of soil /bulk volume

Determination of Soil Density and Porosity

The bulk volume includes the volume of solid particles plus the enclosed pore spaces, that occur in a soil ped or soil clod. The unit measurement of soil weight is the gram, and that of the bulk volume is the cubic centimeter (cc, cm^3, or mL). Hence, the unit of bulk density is g/cc or Mg/m^3. Since the volume of the soil is affected by its water content, the bulk density can be measured on an oven-dry basis or on the weight of a moist bulk volume of soil. The following subscripts are used to indicate the moisture condition of the soil at which the bulk density is measured. Thus, BD_m = bulk density of a moist soil, BD_d = bulk density based on an oven dry soil, and $BD_{1/3}$ = bulk density of soil at 0.3 bar tension (Soil Survey Staff, 1972). Of the three, BD_d, or plain BD, is the most commonly employed term.

In the United States, many people prefer to use lbs/acre to indicate soil mass. The term *acre-furrow-slice* has been introduced in this context, and is defined as the weight (mass) of soil one acre in extent at a normal plow depth (4 - 6 inches) deep. The value depends on the BD and averages 2,000,000 lbs. In order to calculate the acre-furrow-slice, the bulk density value in g/cc must be converted into lbs/cu foot by multiplying by 62.4.

Particle density is defined as the mass (weight) of a unit volume of particles only. Therefore:

$$\text{Particle density} = \frac{\text{weight of soil solids (oven-dry)}}{\text{volume of solid particles}}$$

As previously discussed, the weight of solid particles is measured in grams, whereas the volume of solid particles is measured in cc or cm^3. Hence, the unit for particle density is g/cc or Mg/m^3.

Soil density is related to specific gravity. *Specific gravity* is defined as the ratio of a unit weight of a substance to the unit weight of water, and is, therefore, a dimensionless figure. The unit weight (density) of water is usually taken as 1 g/cm^3 or 1 Mg/m^3 (at 4°C). Specific gravity can be distinguished into (1) apparent specific gravity and (2) true specific gravity. *Apparent specific gravity* takes into account the total soil volume, including pore spaces. *True specific gravity* involves only the mass of solid particles (excluding pore spaces). Consequently, apparent specific gravity is similar to bulk density, whereas true specific gravity is closely related to particle density (Spangler and Handy, 1984).

7.2 MEASUREMENT OF BULK DENSITY

Two methods are commonly used to determine soil bulk density: bulk density measurement of disturbed soil, and bulk density measurement of undisturbed soil. The second method is frequently called the *clod method*.

7.2.1 Bulk Density Measurement of Disturbed Soil

This method is very useful when sampling of soil clods or soil peds is not feasible. Sandy soils are often too loose and friable and do not develop stable aggregates. Similarly soils in nurseries and greenhouses are loose and very friable and do not contain peds or clods. In these cases, bulk density can be measured only by this method.

Procedure

Fill a preweighed 100 mL graduated cylinder with soil. Compact the soil by tapping the bottom of the cylinder ten times with the palm of your hand. Keep adding soil and tapping the cylinder until a tapped soil volume of 100 mL is obtained. Weigh the cylinder containing the soil.

Determination of Soil Density and Porosity

Determine separately the moisture content of the soil sample, and calculate the ovendry weight of the 100 mL soil above.

Calculation

Bulk density = oven-dry weight of 100 mL soil/100 (g/cc)

Range of Bulk Density Values

Low bulk density values (1 - 1.5 g/cc) generally indicate a favorable physical condition of soils for plant growth. The soil has a good structure and many pore spaces for an optimum balance of air and water contents. In natural conditions, low bulk density values are detected in surface soils high in organic matter content. The bulk density often increases with increasing depth in the soil profile.

High bulk density values (1.8 - 2.0 g/cc) indicate a poor physical condition of soils for plant growth. These soils are usually compacted and contain relatively few pore spaces. In natural conditions, high bulk density values are found in subsoils.

Plowing and tillage operations loosen the soil and increase pore spaces, which consequently should decrease bulk density. However, tractor tracts produce compaction, and therefore increase bulk density.

Generally, coarse textured soils exhibit higher bulk density values than do finer textured soils. Sands and sandy loam soils may vary in bulk density from 1.2 to 1.8 g/cc, whereas silt loam, clay loam, and clay soils may have bulk densities ranging from 1.0 to 1.6 g/cc.

7.2.2 Clod Method

This method is used to determine bulk density in an undisturbed soil sample, e.g., soil clods or soil peds (Soil Survey Staff, 1972; Blake and Hartge, 1986). Prior to analysis, the soil clod is coated with a saran resin.

Reagents

A resin, such as Dow Saran F310 resin, is dissolved in acetone or in methyl ethyl ketone at a saran: solvent ratio of 1:4 to 1:8, depending upon the porosity of the clod to be coated. A 1:4 ratio is used for soils that have large pores, but a 1:7 or 1:8 ratio is sufficient for most soil samples. Methyl ethyl ketone is preferred as a solvent over acetone, because it is less soluble in water. Additionally, this solvent exhibits less penetration into a moist clod than does acetone. Both the solvents are flammable and explosive and should be handled with care in a fume hood. After the mixing procedure, the containers should be kept tightly closed to prevent evaporation.

Procedure

Carefully collect soil clods from a soil profile. If roots are present, cut them carefully with a scissor. Tie and hold the clod with a fine copper wire and weigh it. Dip the clod in the saran solution. Hang it up to dry, which usually requires from 15 to 30 minutes. Additional saran coatings can be applied in the laboratory to make the clods more waterproof and to prevent them from bursting during wetting in the analysis. Weigh the coated clod and wire. To determine the volume, weigh the clod while it is suspended in water with a balance that can accept the clod hanging on the balance beam by the thin copper wire. Determine the water temperature to be used for the determination of the density of water. The latter measurement is needed if a correction is required with regard to the density of water.

Determination of Soil Density and Porosity

After the analysis is completed, break the clod. Take a sample for the analysis of soil moisture content, which is needed in order to obtain the oven-dry weight of soil used in the calculations.

Calculation

Bulk density = $W_d/(W_a - n_w \cdot W_w)$

in which W_d = oven dry weight of clod, W_a = weight of saran coated clod (+ wire) in air, W_w = weight of coated clod (+ wire) in water, and n_w = density of water. The density of water is often taken as 1 so that this factor is frequently ignored.

According to Tisdall (1951), the clod method may yield higher bulk density values than other methods.

7.3 MEASUREMENT OF PARTICLE DENSITY

Two methods for the determination of particle density will be discussed. In the first method a graduated cylinder is used; in the second method a volumetric flask is employed. Both analyses are very simple and rapid.

7.3.1 Method Using a Graduated Cylinder

Procedure

Weigh 40 g of soil in a 100-mL graduated cylinder. Measure separately 50 mL of distilled water. Add this amount of water carefully to the soil in the cylinder so that the inner walls of the cylinder are washed clean from soil material. Stir thoroughly with a stirring rod to displace the air, and rinse the stirring rod and the walls of the cylinder with a measured volume (10 mL) of water. Allow the mixture to stand for 5 minutes and record the volume of soil + 60 mL water in the cylinder. This volume minus the total amount of water added equals the volume of water displaced by soil, or the volume of the soil solids.

Determine separately the moisture content of the soil sample by the gravimetric method. This amount of water must be added to the 60 mL used to obtain the total amount of water used.

Calculation

$$\text{Particle density} = \frac{\text{oven dry weight of soil (g)}}{\text{volume of H}_2\text{O displaced by soil (mL)}}$$

Range of Particle Density Values

The particle density of mineral soils has a very limited range, although the density of the individual soil minerals varies considerably. The particle density of most soils varies from 2.60 to 2.75 g/cc. The average particle density of 2.67 is commonly known as the specific gravity of soils.

Quartz, feldspar, and colloidal silicates constitute the major portion of the soil minerals, and their densities fall within the above range. When unusual amounts of heavy minerals, such as magnetite, zircon, tourmaline, and hornblende are present, the particle density value may exceed 2.75 g/cc.

Size and arrangement of soil particles do not affect particle density. However, organic matter, which weighs much less than an equal volume of mineral solids, will decrease the particle density of soils. Some surface soils with high organic matter contents may exhibit particle density values of 2.4 g/cc. Organic

Determination of Soil Density and Porosity

soils (peat and mucks) have extremely low particle densities.

7.3.2 Method Using a Volumetric Flask

In principle, this method is similar to the pycnometer method for the determination of specific gravity. However, larger amounts of soil are used with the volumetric flask than with the pycnometer method. The use of larger amounts of soil samples results in more accuracy in weighing.

Procedure

Fill a preweighed 100-mL volumetric flask with cool, boiled distilled water to the mark. Record the weight and discard the water. Dry the flask thoroughly. Weigh 50 mg of soil and transfer this amount of soil quantitatively in the volumetric flask. Use a pipette to measure 50 mL of water and add this water to the soil in the volumetric flask, carefully washing the adhering soil from the neck of the flask at the same time. Place the flask on a hot plate and heat gently until the water starts to boil. Boiling is necessary to displace air from the soil. Cool the content and make up the volume to 100 mL with boiled distilled water at room temperature. Weigh the volumetric flask with its contents and determine the temperature of the content.

Determine separately the moisture content of the soil sample by the gravimetric method for calculations on an oven-dry basis.

Calculation

$$\text{Particle density} = \frac{\text{oven-dry weight of soil}}{\text{g (soil+water)} - \text{g soil}}$$

In this equation, the weight of soil + water = weight of the flask (with soil + water) minus the weight of the flask only, whereas the weight of the soil = the weight of the flask (with soil + water) minus the weight of the flask (+ water). The volume of the solid particles is then the amount of water displaced by the soil solids (1 g of water = 1 cc of water). In the calculations above, the density of water is assumed to be 1. If corrections are needed, the density of water, n_w, should be used as a correction factor in the denominator.

7.4 DETERMINATION OF PORE SPACES

Pore space is that portion of the soil volume not occupied by solid particles, but occupied by air and water. In micromorphology the term *void* is used for the term for pore space (Bullock et al., 1985). The arrangement of solid particles determines the amount of pore space or total porosity of soils. Sandy surface soils have a total porosity ranging from 35% to 50%, whereas finer textured soils exhibit pore spaces ranging from 40% to 60%. Thus, finer textured (silty and clayey) soils have a larger total porosity than do coarser (sandy) soils. Compacted subsoils may have pore spaces as low as 25% to 30%.

Two types of individual pore spaces occur in soils: macropores and micropores. There is no sharp line of demarcation between these two types of pores. Macro-(large) pores function in air and water movement, whereas micro-(small) pores retain or hold soil moisture. Macropores dominate in sandy soils. Thus, air and water move fairly rapid in these soils in spite of the small amount of total pore space. Fine textured soils have a preponderance of micropores. These soils have a greater water holding capacity as compared to sandy soils, but the movement of air and water is comparatively more restricted.

Determination of Soil Density and Porosity

Procedure

The simplest method to determine pore space is through the measurements of bulk density and particle density (Tan, 1990). From the results of the two measurements, the percentage of pore space can be calculated by using the following equation:

$$\% \text{ Pore space} = 100 \times \frac{\text{particle density - bulk density}}{\text{particle density}}$$

Examples of Calculations

1. A sample of soil weighs 220 g and contains 15% H_2O. The volume of the soil sample is 150 mL. Calculate the bulk density.

    Oven-dry weight of soil = 220 x 100/100 + 15 = 191.3 g

    Bulk density = 191.3/150 = 1.3 g/cc

2. The soil sample above displaces 73.6 mL of water. Calculate the particle density.

    Particle density = 191.3/73.6 = 2.6 g/cc

3. Calculate the percentage of pore space.

    % Pore space = 100 x [(2.6 - 1.3)/2.6] = 50%

CHAPTER 8

SOIL pH MEASUREMENT

8.1 DEFINITION AND PRINCIPLES

The degree of acidity or alkalinity in soils, also known as *soil reaction*, is determined by the hydrogen ion (H^+) concentration in the soil solution. An acid soils has more H^+ than OH^- ions, whereas a basic or alkaline soil contains more OH^- than H^+ ions. To characterize these conditions, the term soil pH is used. The term pH was introduced by Sörensen in 1909 (Tan, 1982), and is defined as:

$$pH = -\log\,(H^+)$$

or

$$pH = \log\,1/(H^+)$$

In these equations, p refers to the negative logarithm and H^+ to the *free* H^+ ion concentration. The unit of (H^+) is measured in activity or in moles/L. *Activity*, also known as the effective concentration, is that part of the *actual* H^+ ion concentration which participates in chemical reactions. In a dilute solution,

Soil pH Measurement

such as a soil solution, activity = actual concentration. The preferred unit for concentration is moles/L, but since 1 mole of H^+ = 1 g, grams/L is accepted in the calculation of pH values.

The values of H^+ ion concentration are usually very small. In a neutral solution, which contains equal amounts of H^+ and OH^- ions, the H^+ ion concentration is 0.0000001 g/L. The use of these small numbers is very confusing; therefore, it is more convenient to use pH values. In aqueous solutions, the pH scale ranges from 0 to 14. The higher the pH value, the lower the H^+ ion concentrations, or, in other words, the less acid the solution, hence the higher the OH^- ion concentration. For example, at a H^+ ion concentration of 0.001 g/L, the pH = - log 0.001 = - (-3) = 3. At a H^+ ion concentration of 0.000001 g/L, the pH = - log 0.000001 = - (-6) = 6.

The mass action law states that the product of the concentration of H^+ and OH^- ions is always constant. This law may be written as an equation:

$$(H^+)(OH^-) = 10^{-14}$$

By taking the negative logarithms on both sides, the equation becomes:

$$pH + pOH = 14$$

Since the sum of pH and pOH is constant, the concentration of H^+ and OH^- ions are interrelated, and vary inversely. Thus, only one ion, needs to be determined, the H^+ ion, in order to know the other.

The H^+ ions may be present in soils as adsorbed H^+ ions on the surface of the colloidal complex, or as free H^+ ions in the soil solution. The adsorbed H⁺ ions create the *reserve acidity*, also called the *potential* or *exchange acidity* of soils. The free H⁺ ions are the reason for the so-called *active acidity* of soils. Soil pH refers to this active acidity. Taken together, the active plus potential acidities are called the *total acidity*. The reaction of acid soils can be influenced only when enough lime is added to react with the total acidity (active + reserve acidity). The greater the cation exchange capacity of the soil, the greater will be the reserve acidity, and the more difficult to reduce the total acidity. This resistance to change in soil reaction in general is called the *buffer capacity*.

Based on soil pH values, several types of soil reactions are distinguished as follows (Tan, 1982; Brady, 1990):

	pH		pH
Slightly acid	7.0 - 6.0	Slightly alkaline	7.0 - 8.0
Moderately acid	6.0 - 5.0	Moderately alkaline	8.0 - 9.0
Strongly acid	5.0 - 4.0	Strongly alkaline	9.0 - 10.0
Very strongly acid	4.0 - 3.0	Very strongly alkaline	10.0 - 11.0

Most plants grow best in soils with a slightly acid reaction. In this pH range, nearly all plant nutrients are available in optimal amounts for plant growth. Soils with a pH below 6.0 are more likely to be deficient in some available nutrients. Calcium, Mg and K are especially deficient in acid soils. In strongly and very strongly acid soils, Al, Fe, and Mn may exist in toxic quantities (very high amounts) because of their increased solubilities. If phosphates are present, these elements will react with the phosphates to form insoluble phosphates. In alkaline soils (pH >7.0), Fe, Mn, Zn, and Cu become unavailable for plant growth. Many other soil properties and processes are

Soil pH Measurement

affected by soil pH, e.g., clay mineral formation and microbial activity. Because of the above, Jackson (1956) indicated that soil pH was perhaps the most important soil chemical property. Besides its importance in agriculture, pH is also an important factor in such industries as the paper industry, rubber industry, sugar industry, medicine, and food and textile processing. In the paper industry, the pH level of the raw pulp affects the quality of the paper as to how it prints and takes ink. In rubber factories, pH plays a significant role in the polymerization and coagulation process. In medicine, pH is used to show the acid-base equilibrium of blood as an indicator of general health. It also finds application in the determination of acidity in gums and stomach fluid in the control of gum disease and ulcers, respectively. Dying and bleaching process in the textile industry are dependent on pH. In the food industry, pH measurements are performed to determine spoiling of meat and other food products. Therefore, because pH is an important factor in so many fields, it is essential to know how to measure pH properly.

In general, two types of pH measurements in soils can be distinguished: the indicator or colorimetric method and the potentiometric method.

8.2 INDICATOR OR COLORIMETRIC METHOD

The basic requirement of most indicator methods for pH determination is a clear solution extract. This requirement necessitates the use of wide soil-water ratios, which are not comparable to natural soil conditions.

The method makes use of indicators, and is applied in the field as a rapid test for soil pH. Indicators are usually high molecular weight, weakly dissociated organic acids, or bases. In this method, in which color is used to indicate pH levels, the free ion of the indicator has a color different from the dissociated molecule. The equilibrium concentration between the dissociated molecule and the undissociated indicator governs the color. The dissociation reaction is influenced by H^+ ions and can be represented as:

$$HI \rightleftharpoons H^+ + I^-$$

in which HI = the undissociated indicator with the acid color, and I^- = the dissociated indicator with the alkaline color. According to the mass action law,

$$[(H^+)(I^-)]/HI = k$$

This equation may be rearranged as:

$$(I^-)/(HI) = k/(H^+)$$

or it may be converted into a log equation:

$$\log (I^-)/(HI) = \log k/(H^+)$$
$$= \log k - \log (H^+)$$

If $(I^-)/(HI) = 1$, then: $\log (I^-)/(HI) = 0$ or

$$\log k - \log (H^+) = 0$$
$$-\log k = -\log (H^+)$$

Soil pH Measurement

pk = pH

The factor -log k or pK is called the *indicator constant*.

The point at which pK = pH is a critical point, and the pH is called the *critical pH* (Jackson, 1958). A slight change in concentration of the dissociated and undissociated molecules from this critical point produces a pronounced color change. This change in color is used to determine the soil pH. The critical pH value differs from indicator to indicator (Table 8.1). Generally, a mixture of selected indicators is used in order to measure the soil pH from 0 to 14. Several mixed indicators are commercially available under different names, e.g., *universal indicator*, and *duplex indicator*. The pH scale of the duplex indicator (Lamotte Chemical Products, Towson, Baltimore), ranging from pH 3 to 11, is indicated by a color chart from red to blue. Red colors indicate an acid reaction, and yellow to light green colors indicate a slightly acid, neutral to slightly basic reaction. Blue colors indicate a basic reaction.

Reagent

Duplex indicator, or if desired, a mixed indicator, measuring pH levels from 2.0 to 10.0, can be prepared by dissolving in 100 mL ethanol:

80 mg bromthymol blue
40 mg methyl red
60 mg methyl yellow
20 mg phenolphthalein
100 mg thymol blue

Titrate the mixture with 0.1 M NaOH solution until yellow. The pH level corresponds to the colors listed in Table 8.2.

Table 8.1 Some pH indicators commonly used in soil analysis (Jackson, 1958; Kolthoff and Sandell, 1952)

Indicator	Critical pH	Color change[1]
Thymol blue[2]	1.9	R-O-Y
Dinitrophenol	3.1	C-Y-Y
Methyl orange	3.7	R-Y-O
Brom phenol blue	4.0	Y-Pu-V
Brom cresol green	4.6	Y-G-B
Chlor phenol red	5.6	Y-OP-V
Methyl red	5.7	R-O-Y
Brom thymol blue	6.9	Y-G-B
Phenol red	7.3	Y-RO-V
Phenolphthalein	8.3	C-P-P
Thymolphtalein	9.4	C-B-B
Alizarin yellow R	10.3	Y-O-R

[1] Color at center is at critical pH; B = blue, C = colorless, G = green, O = orange, P = pink, Pu = purple, R = red, V = violet, Y = yellow.
[2] Thymol blue has two critical pH values.

Procedure

1. *Laboratory Procedure* Weigh 15.0 g of soil in a clean 100-mL centrifuge tube. Add 30 mL of distilled water, and swirl frequently for 5 minutes. Centrifuge the mixture at 2500 rpm for 15 minutes, and filter the supernatant solution into a 150 mL beaker. Transfer 10 mL of the clear solution into a test tube. Add 5-10 drops of duplex indicator or mixed indicator, and swirl to develop the color. Match the color with the color chart to determine the pH.

   If the mixed indicator is used, prepare a series of standards, and match the color with these standards.

Table 8.2 The pH and Corresponding Color of Mixed Indicator

pH	Color
2.0	Crimson-red
3.0	Red
4.0	Orange-red
5.0	Orange
6.0	Yellow
7.0	Yellowish-green
8.0	Green
9.0	Bluish-green
10.0	Blue

2. *Field Procedure* Place a small sample of soil in a plastic spoon. Wet the sample with distilled water and allow the mixture to react for 5 minutes. Tilt the spoon to separate the liquid from the soil, and add 2 to 3 drops of duplex indicator to the solution. Match the color developed with the color chart to determine the pH.

8.3 POTENTIOMETRIC METHOD

This method uses electrodes to measure the H^+ ion concentration in the soil solution. Although a large variety of electrodes are available, three types are commonly employed in pH measurement. They are the glass electrode, which is the indicator electrode; the calomel electrode, which is the reference electrode; and the combination glass electrode, in which both the indicator and the reference electrodes are combined into one electrode (Figure 8.1). When placed in a solution, an electrical potential difference develops between the indicator and reference electrodes (or the combination electrode) and the solution. The magnitude of the potential difference is proportional to the H^+ ion concentration, and can be converted into pH units. The potential difference, usually denoted as the potential, E, is given by the Nernst equation (Tan, 1982):

$$E = E_o + RT/nF \, (\ln a_m)$$

In this equation, E_o = standard potential, R = gas constant, T = degrees Kelvin, n = valence, F = Faraday constant, a = activity, and m = ion with valence n.

Changing ln into log yields the following equation:

$$E = E_o + (RT/nF) \; 2.3 \log a_m$$

Since 2.3 RT/F = 0.059, the equation can be changed into:

$$E = E_o + 0.059/n \, (\log a_m)$$

In pH measurements $a_m = H^+$, and n = 1. Hence, the equation becomes:

$$E = E_o + 0.059 \log (H^+)$$

$$= E_o - 0.059 pH$$

This equation indicates that if the pH changes with 1 unit, the potential (voltage) changes 0.059 V.

Soil pH Measurement

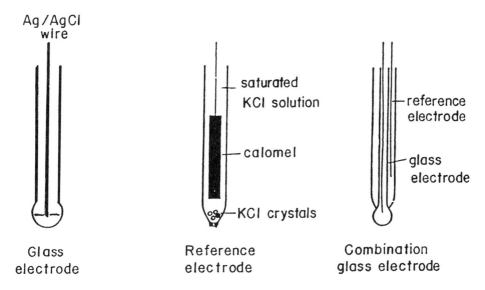

Figure 8.1 Examples of glass electrodes for pH measurement.

Reagents
Commercial buffer solutions of pH 7.0 and pH 4.0 for calibration of the pH meter.

Procedure
Weigh 25 g of soil in a clean beaker. Add 25 mL distilled water and stir or swirl frequently for 15 minutes.

Calibrate the pH meter prior to use by inserting the combination glass electrode in a buffer solution of pH 7.0.

Adjust the pH meter to read pH 7.0. Rinse the electrode with distilled water and place it in a buffer solution of pH 4.0 to read pH 4.0. Rinse the electrode again with distilled water and place it in the soil suspension above. Read the pH in the scale of the pH meter.

After measuring the pH, rinse the electrode with distilled water and place it in the buffer solution of pH = 7.0. The glass electrode requires a hydrated layer on the outer glass wall to respond accurately to H^+ activity. If the electrode is allowed to dry, its function is impaired. Therefore, the glass electrode must be soaked in water, preferably in a buffer solution of pH 7.0, for storage.

8.4 FACTORS AFFECTING pH MEASUREMENTS

8.4.1 Suspension Effect

The pH of a soil suspension (soil and water mixed thoroughly) is usually lower than the pH of the supernatant produced when the water is separated from the soil by centrifugation or gravitational forces. The H^+ concentration is higher at the clay surfaces than in the bulk solution. In a soil suspension, the electrode registers the H^+ ions both in the solution and at the surfaces of the clay. When pH measurements are performed in the supernatant solution, the electrode measures only the H^+ ion concentration of the bulk solution. The difference in results is called the suspension effect.

8.4.2 Dilution Effect

Soil pH can be measured at different soil:water ratios, e.g., at soil:water ratios of 1:1, 1:2, 1:5, or 1:10, and different results are usually obtained. The soil pH tends to be higher when measured at wider (higher) soil:water ratios (Jackson, 1958). When more water is used, the concentration of H^+ ions becomes diluted, and the measurement yields higher pH values. Most pH determinations are performed in a suspension with a soil:water ratio of 1:1 or 1:2, since this condition is closer to natural conditions than the wider ratios. A soil:water ratio of 1:10 is sometimes used for specific purposes, and the resulting pH is called a *hydrolytic pH* (McGeorge, 1944). The hydrolytic pH increases in value with the degree of Na-saturation.

A soil:water ratio narrower than 1:1 is used by Chapman et al. (1940) for pH measurement at the *sticky point*. In this procedure, the soil sample is carefully wetted with distilled water until the sticky point is reached. At this point, the soil looks like a thick paste. If a spatula is pressed on the soil paste, it will stick to the spatula. A glass electrode is then inserted carefully in the soil paste, and the pH measured in the usual way. The sticky point method in pH measurements can be applied in soils containing appreciable amounts of clay, since it is difficult to form a soil paste with sands. Applying this method in sandy soils may cause damage to the glass electrode.

8.4.3 Sodium Effect

The sodium effect occurs in the pH measurement of alkaline solutions. At high pH values, hydrolysis of the glass membrane and/or Na ions takes place, causing a lowering of the pH observed in the range of pH >9. To reduce the error caused by this so-called sodium effect, the electrode is often standardized with a buffer solution containing the same amount of Na as the test solution.

8.5 WATER pH VERSUS BUFFER pH

To decrease fluctuation in pH readings during analysis, many scientists perform the analysis also in 0.01 M $CaCl_2$ or 1 M KCl solutions, instead of in water only (Schofield and Taylor, 1955; Puri and Asghar, 1937). The procedure is similar to the determination of pH in water as discussed above. The pH measured in these solutions is called the *buffer pH*. The use of $CaCl_2$ or KCl solutions has the advantage of (a) decreasing the effect of the junction potential of the calomel reference electrode, (b) preventing dispersion of the soil, and (c) equalizing the salt content of soils. The *liquid junction potential* is present when the glass-calomel electrode pair is used, but it is now a minor problem due to the use of a saturated KCl-AgCl filling for the calomel electrode. The variation in salt content of soils is believed to be a more serious problem, since it affects the ionic activity in soils, which may cause the pH reading to vary considerably. This error is of special importance in pH measurements of dilute solutions. Therefore, the soil sample is swamped with $CaCl_2$ or KCl, and it is thought that the readings will produce more constant pH values than measurements in water. Usually the value of the buffer pH is lower than the value of pH in water.

The question is now whether the buffer pH represents the true pH value of the soil under investigation. The significance of the buffer pH also depends on the willingness of the scientist to accept a 0.01 M $CaCl_2$ or a 0.1 M KCl soil solution as a natural (true) soil solution. Arguments can be presented from both sides. Unpublished results of investigations conducted at the University of Georgia with a Tifton soil (Fine-loamy, siliceous, thermic Plinthic Kandiudults) in south Georgia indicate that over a period of three years, pH measurements in water and KCl produce more stable values than do determinations in $CaCl_2$ (Figure 8.2). The soil is limed on day 1, and the pH starts to increase after 42 days. When measured in a soil water suspension, the pH increased slightly from 112 to 583 days and starts to decrease after 583 days. Measured in a $CaCl_2$ suspension, the soil pH starts to increase after 42 days, and it continues to increase considerably until 239 days. The pH then decreases slightly, and after 583 days it decreases rapidly. The soil pH values in KCl solutions are almost similar to the soil pH values in water and was more stable from 42 to 758 days (G.J. Gascho, personal communication), indicating that measurements in KCl yield a more reliable buffer pH than do measurements in $CaCl_2$.

8.6 DETERMINATION OF LIME REQUIREMENT

As indicated above, crops yield best in soils with pH levels between 6.0 and 7.0. Lime is added to raise the pH of acid soils, and the amount of lime required to raise the pH to 6.0-6.5 is called the lime requirement. A number of methods is available for the determination of lime requirements of acid soils. The Adams and Evans (1962) method, because of its ease and rapidity in analysis, is selected and discussed here.

Reagents

The Adams and Evans buffer is prepared by dissolving in 1125 mL distilled water (under constant stirring): 15.75 g KOH, 30.0 g p-nitrophenol, and 22.5 g H_3BO_3. Heat slightly, if necessary, to dissolve the crystals. Allow the solution to cool, and make up to 1.5 L with distilled water. When this buffer is properly prepared, its pH should be 8.0.

Procedure

Weigh 20 g of soil in a beaker or paper cup. Add 20 mL of distilled water, stir, and allow the suspension to stand for 30 m. Determine the pH with a pH meter, and save the soil suspension. The measured pH is called pH_{water}.

Continue the analysis by adding to the soil suspension above 20 mL of the Adams and Evans buffer. Stir for 5 m.

Calibrate the pH meter to read pH 8.0 with a diluted Adams and Evans buffer (20 mL buffer + 20 mL water). Rinse the electrode, and measure the pH of the soil suspension containing the buffer. This is called the pH_{buffer}.

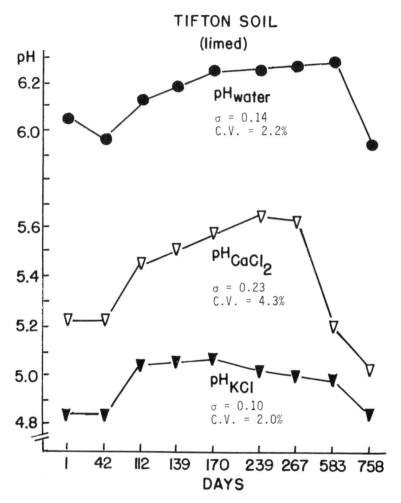

Figure 8.2 Measurement of pH in water versus $CaCl_2$ and KCl solutions in a Tifton soil over a three year period. Courtesy: G. J. Gascho, University of Georgia.

Soil pH Measurement

Table 8.3 Lime Requirements to Raise Soil pH to 6.5. Figures in Table x 1000 lbs Calcitic Limestone per Acre

pH$_{H_2O}$	Buffer pH									
	7.90	7.80	7.70	7.60	7.50	7.40	7.30	7.20	7.10	7.00
6.4	0	0	0	0	0	1	1	1	1	1
6.3	0	0	0	1	1	1	2	2	2	2
6.2	0	0	1	1	1	2	2	2	3	3
6.1	0	0	1	1	2	2	2	3	3	3
6.0	0	1	1	2	2	2	3	3	4	4
5.9	0	2	2	2	2	3	3	4	4	5
5.8	0	2	2	2	2	3	4	4	5	5
5.7	0	2	2	2	3	3	4	4	5	5
5.6	0	2	2	2	3	3	4	5	5	6
5.5	1	2	2	2	3	4	4	5	5	6
5.4	2	2	2	2	3	4	4	5	6	7
5.3	2	2	2	3	3	4	5	5	6	7
5.2	2	2	2	3	3	4	5	6	6	7
5.1	2	2	2	3	4	4	5	6	7	7
5.0	2	2	2	3	4	4	5	6	7	8
4.9	2	2	2	3	4	5	5	6	7	8
4.8	2	2	2	3	4	5	6	6	7	8
4.7	2	2	2	3	4	5	6	7	7	8
4.6	2	2	2	3	4	5	6	7	8	8
4.5	2	2	3	3	4	5	6	7	8	9

Determine the amount of lime ($CaCO_3$) required to bring up soil pH to 6.5 from Table 8.3.

8.7 DETERMINATION OF LIME POTENTIAL

As discussed previously, pH measurements are affected by the suspension effect and by the activity of electrolytes, especially when the soil is limed with Ca^{2+} and Mg^{2+}. These may cause errors in soil pH readings. To control the errors, Schofield and Taylor (1955) suggest the use of a correction factor by considering the Ca^{2+} concentration in the calculation of pH. This is done by using the activity ratio of $(H^+)/(Ca^{2+})^{0.5}$, which is constant. Using the negative logarithm for this function yields:

$$-\log (H^+)/(Ca^{2+})^{0.5} = -\log k$$

$$pH - 0.5pCa = pK$$

The function *pH - 0.5pCa* is called the *lime potential*. Its value is believed to be less dependent on salt concentrations, and thus it is a better index of the lime requirement.

Reagent

0.01 M $CaCl_2$ solution. Weigh 1.11 g $CaCl_2$ in a 1 L volumetric flask. Add approximately 500 mL distilled water. Dissolve it under constant stirring, and make up to volume with distilled water.

Procedure

Weigh 20 g of soil in a beaker, and add 40 mL of the $CaCl_2$ solution. Stir for 30 m and measure the pH of the suspension. Swirl before measurement.

Soil pH Measurement

Calculation

Lime potential = pH - 0.5pCa

When 0.01 M $CaCl_2$ is used, the activity of Ca^{2+} is known and constant. At 0.01 M, 0.5pCa = 1.14, and the equation above changes into:

Lime potential = pH - 1.14

CHAPTER 9

CATION EXCHANGE CAPACITY DETERMINATION

9.1 DEFINITION AND PRINCIPLES

Clay and humus in soils are colloidal particles with large surface areas. Most of the clay particles are electronegatively charged. However, humic and fulvic acids, the chemically active fractions of humus, may be positively or negatively charged. The negative charges are usually neutralized by a swarm of positively charged ions (cations). These cations, which are held electrostatically on the surface of the soil colloids, are called *adsorbed cations*. Adsorbed cations can be replaced or exchanged by other cations from the soil solution. The process of replacement is known as *cation exchange*, and the ions involved in the exchange reaction are called *exchangeable cations*. The term *cation exchange* is preferred over the term *base exchange*, since H^+, an important cation in the exchange process, is not a base. The rate of the reaction is virtually instantaneous. To maintain electroneutrality in soils, exchange reactions occur in equivalent amounts, as illustrated below:

$$\text{micelle-Ca} + 2H^+ \rightleftharpoons 2H\text{-micelle} + Ca^{2+}$$

Cation Exchange Capacity Determination

The capacity of soils to hold (adsorb) and exchange cations is, therefore, called *cation exchange capacity*. The cation exchange capacity, frequently abbreviated as *CEC*, is a quantitative measure of all the cations adsorbed on the surface of the soil colloids. The amount is usually expressed in milliequivalents (me) per 100 g of soils. In SI units, this amount equals cmol(+)/kg of soil. Hence, the CEC can be formulated as:

$$\text{CEC} = \Sigma \text{ exchangeable cations(me)}/ 100 \text{ g soil}$$
$$= \Sigma \text{ exchangeable cations\{cmol(+)\}/kg soil}$$

The USDA, Soil Survey Division, uses me/100 g of clay as a unit.

Since cation exchange reactions are functions of the surface area and the amount of negative charge on the surface, the CEC is proportional to the specific surface (s) and to the surface charge density (σ):

$$\text{CEC} = s \times \sigma$$

The surface charge density is expressed as:

$$\sigma = \sqrt{[2\varepsilon\eta KT/\pi]} \ \sinh(ze\psi/2KT)$$

where ε = dielectric constant, η = electrolyte concentration in numbers of ions per mL, K = the Boltzmann constant, T = absolute temperature in degrees Kelvin, z = valence, e = electron charge in esu, and ψ = surface potential in statvolts (Tan, 1982).

Because of differences in concepts, and in view of the presence of permanent and variable charges contributing to cation exchange reactions, several types of CEC have currently been identified. The most important of these types are:

1. $CEC_{NH\ OAc}$, which is the CEC obtained by the NH_4OAc method
2. Effective CEC, or ECEC, which equals the exchangeable bases (extracted by the NH_4OAc method) + extractable Al (in 1 M KCl)
3. CEC_p, which is the CEC produced by the permanent negative charges of the clay minerals
4. CEC_v, which is the CEC resulting from the pH dependent (or variable) charges of the inorganic and organic colloids
5. CEC_t, which is the total CEC caused by both the permanent and variable charges of the soil colloids.

Adsorption and cation exchange are of great practical significance in soil physics, soil chemistry, soil fertility, nutrient retention in soils, nutrient uptake by plants, lime and fertilizer application. Adsorbed cations are generally available to plants by exchange with H^+ ions, which are generated by the respiration of plant roots. Nutrients added to the soil, in the form of fertilizers, are retained by the colloidal surfaces. Cations that may pollute the groundwater may be filtered by the adsorptive action of the soil colloids. As such, the adsorption complex (clay and humus) serves the soil as a storage and buffering capacity for cations. Furthermore, the cation exchange capacity helps to determine the amount of lime to be applied to acid soils.

The type of exchangeable cations may also affect the soil physical properties. Calcium and Mg in exchangeable forms are good sources for promoting good soil structure and good soil

Cation Exchange Capacity Determination

tilth. On the other hand, the presence of large amounts of Na in exchangeable form may have a harmful effect on soil structure. The presence of Na may cause the soil to disperse with the consequent destruction of soil structure and pore spaces. A dispersed soil is sticky and plastic when wet. When dry, a dispersed soil is massive and hard, hence is much less permeable to water and air. The concentration and distribution of exchangeable cations in a soil pedon can also be used as a measure of the degree and rate of weathering and leaching.

9.2 FACTORS AFFECTING CEC DETERMINATION

9.2.1 pH of the Leaching (Displacing) Solution

No general agreement is present as to the pH of the leaching solution for the exchange of the natural soil cations. Although many people believe that the leaching solution should have a pH as close as possible to the pH of the soil under investigation, a pH of 7.0 is most commonly employed, especially when NH_4OAc is used as the leaching solution. However, when exchangeable H^+ ions are to be included in the CEC determination, a solution with a pH ranging from 8.0 to 8.5 must be used if there is to be an effective replacement of adsorbed H^+ ions. The pH and type of salt used have relatively little effect upon the exchange of the bases, and the sum of exchangeable bases ($Na^+ + K^+ + Ca^{2+} + Mg^{2+}$) will not differ greatly if the replacement solution is NaOAc, KOAc or Ba-salts at a pH of 7.0 or 8.5. However, the exchangeable H^+ ion concentration extracted is distinctly different when the exchange reaction is performed at a pH of 7.0 or 8.5. A leaching solution with a higher pH usually results in an increased displacement of H^+ ions. Since the CEC is by definition the sum of exchangeable (bases + H^+), an analysis detecting higher amounts of exchangeable H^+ ions will, therefore, yield a higher CEC value. The magnitude of this pH effect depends upon the nature of the exchange complex, the nature and concentration of the leaching solution, and the analytical approach. These factors will be discussed in the following sections.

9.2.2 Nature of the Exchange Complex

Six distinct types of soil colloids are involved in cation exchange reactions. They are: the amorphous and paracrystalline clays, 1:1 lattice type of clays (kaolinite group), 2:1 lattice type of clays (smectite group), 2:2 lattice type of clays (chlorite group), sesquioxide clays (Al- and Fe-oxide clays), and the organic colloids (humic and fulvic acids).

The CEC values and the behavior of these six categories of soil colloids in cation exchange reactions - in particular in exchangeable H^+ reaction - are distinctly different. The bonding of H^+ in 1:1 types of clay is different from the bonding of 2:1 lattice types of clays. For example, in kaolinite, the proton is present as -OH groups on the surface planes of the mineral. These -OH groups are called *exposed hydroxyl* groups and will dissociate their protons at high soil pH values. They are the sources for the variable negative charges, because the dissociation of the H^+ ions from the OH groups is pH dependent. Exposed hydroxyl groups are also present in the sesquioxidic, amorphous, and paracrystalline clays. The OH groups in the smectitic clays are not present as exposed OH groups. They are contained within the crystal, covered by layers of silica tetrahedrons. Therefore, smectite has its negative charge mainly from isomorphous substitution. Because this charge is independent of pH, it is, therefore, called a *permanent charge*. On the other hand, humic and fulvic acids are amphoteric compounds and have different forms of dissociable H^+, e.g., -R-COOH, phenolic-OH and R-NH. The proton from the carboxyl (COOH) group will dissociate at pH 3.0, whereas the proton of the phenolic-OH group starts to dissociate at pH 9.0. Consequently, organic colloids have variable negative charges.

The magnitude of negative charges will, therefore, be different from colloid to colloid. The organic colloids are usually highest in total negative charges, whereas smectite is known to to possess lower amounts of total negative charges. Since these charges are the reasons for the cation exchange reactions, consequently the organic colloids exhibit a higher and smectite a lower CEC value (Table 9.1).

Table 9.1 Cation Exchange Capacity of Soil Inorganic and Organic Colloids (Tan, 1993; Brady, 1990)

Colloid	CEC	Attributable to	
		Variable charge	Permanent charge
	cmol(+)/kg	-------%-------	
Organic (humus)	200	90	10
Vermiculite	150	5	95
Smectite	100	5	95
Illite	30	20	80
Chlorite	30	20	80
Kaolinite	8	95	5
Gibbsite	4	100	0
Goethite	4	100	0

9.2.3 Nature and Concentration of the Leaching Solution

The type of the cations used in the leaching solution affects the results of analysis. The cations, introduced as exchange material, are subject to adsorption. The rate of adsorption depends upon the valence and size of the cations. In general, monovalent ions tend to be adsorbed in smaller amounts than divalent ions. On the other hand, the smaller the size of the cation, the better the rate of adsorption. For monovalent cations, the following lyotropic series is reported on the basis of size (Tan, 1982):

$$H < Cs < Rb < K < Na < Li$$

The cations above are listed in increasing order of hydrodynamic radius of the atoms. The hydrogen ion has the smallest hydrodynamic radius, hence its rate of adsorption is greater than that of Li. As can also be noticed, K is smaller in size than Na, and should, therefore, be adsorbed in larger amounts than Na. Potassium, then, should be a better exchanger than Na, although Na is more widely used in cation exchange determinations. Potassium is subject to fixation by clays, which may interfere in the CEC determination.

In addition, the ability of the cation to displace is not always proportional to the rate of becoming adsorbed. Ideally, the cation of the leaching solution should be adsorbed in amounts equivalent to the amounts of ions displaced. For example, it has been noted that NH_4^+ has a high displacing power, but it is not adsorbed in high amounts. On the other hand, Ba^{2+} is adsorbed in large amounts without displacing equivalent amounts of cations (Golden et al., 1942).

Not all types of cations can be used as exchange material. Generally, only monovalent cations and some divalent ions are usable in CEC determinations. The cations Mg^{2+}, Fe^{2+}, Fe^{3+}, and Al^{3+} are unsuitable because they interfere with the analysis by forming basic salts and insoluble hydroxides (Bower and Truog, 1940). Leaching solutions containing $BaCl_2$ must be used with

Cation Exchange Capacity Determination

caution because the Ba^{2+} ions will react with CO_2 gas to form $BaCO_3$, which is insoluble and hence precipitates. In addition, the reaction to form $BaCO_3$ interferes with the analysis of CEC.

The concentration of the displacing or leaching solution has also a significant influence on CEC determination. In general, the more concentrated the displacing solution, the more H^+ ions will be displaced. Dilute solutions usually yield lower results. The net effect is that the value obtained by an analysis with a strong (concentrated) leaching solution corresponds to one obtained with a dilute solution that has a higher pH value.

9.2.4 Presence of Undesirable Interactions

The fixation of certain ions may yield low CEC values. The use of K as a displacing ion, as has been mentioned above, interferes with the determination of CEC, because K tends to be fixed by soil minerals. Ammonium (NH_4^+) ions are also subject to NH_4-fixation, and consequently may cause a similar interference as K-fixation in the analysis of CEC.

Potassium may be present as fixed-K in the soil under investigation. The *solubility* of this naturally occurring fixed K may introduce a significant error into the result.

Precipitation reactions, as discussed earlier for Ba^{2+}, are also sources for errors. In addition to the reaction with CO_2, Ba^{2+} ions can also react with sulphate originally present in soils to form $BaSO_4$, which is also insoluble.

Hydrolysis of the displacing cations, with the consequent formation of insoluble hydroxides, is another source of error. For example Mg^{2+} ions are reported to have a tendency to hydrolyze in soils. The consequent formation of $Mg(OH)_2$ affects the result (Kelley, 1948).

9.2.5 Analytical Approach

The concept used in the method for the determination of CEC

has an important bearing on the result. Two general concepts are available for the analysis:

1. CEC = sum of exchangeable (bases + H^+). The basic principle is then to displace the natural exchangeable soil cations by a known cation. The displaced ions are measured, and the sum of these cations (in me/100 g) is taken as the CEC of the soil.
2. CEC = amount of the adsorbed fraction of cations used for displacement of the natural cations. In this case, the natural exchangeable cations are also displaced by a known cation. However, the fraction of the introduced cation that is adsorbed by the soil is then re-exchanged by another cation, and its concentration determined and taken as the CEC.

The difference in approach produces different results in CEC. There are two sources for the differences between the two types of concepts. In the first approach, the bases are determined with one method, whereas the H^+ is determined with another (separate) method. In the second approach, the adsorbed amount of the displacing ion serves as the CEC. However, the displacing ion is adsorbed to a greater or lesser extent than the amount of natural cations it displaces. Frequently, the first approach yields higher CEC values than the second, because usually higher H^+ ion concentrations will be measured by the separate method.

9.2.6 Limitations of the Analysis

In addition to normal analytical uncertainty and errors, limitations in the technique may cause other errors. For example, in the determination of CEC by measuring the fraction of the displacing cation adsorbed by the soil, it is necessary to remove the excess of the displacing solution without disturbing (removing) part of the adsorbed fraction. This step is usually

Cation Exchange Capacity Determination

performed by repeated washings with ethanol, and there is a question that the washings, especially with a mixture of 50% ethanol and 50% water, may also remove some of the adsorbed fraction of the displacing cation (Peech, 1945).

9.3 METHODS OF CEC DETERMINATION

As indicated before, the importance of CEC in the many branches of soil science has resulted in the development of many methods for its determination. Some of these methods to determine CEC use NaOAc or $BaCl_2$ as the displacing salts, other methods use NH_4OAc, KOAc, $Ba(OAc)_2$ and $BaCl_2$-triethanolamine. Rhoades (1982) reported the application of NaOAc and $BaCl_2$ for the analysis of CEC in basic and acid soils, respectively. However, the most widely employed method uses NH_4OAc. Moreover the criteria for the placement of soils in the U.S. Soil Taxonomy is based on the CEC determined by the NH_4OAc method. In view of the above, and in view of the current distinctions into effective CEC or ECEC, CEC_p, CEC_v, and CEC_t, as indicated earlier, the NH_4OAc method and the methods for the determination of ECEC, CEC_p, and CEC_t will be discussed in this book. The CEC_v is determined by difference as follows:

$$CEC_v = CEC_t - CEC_p$$

9.3.1 Ammonium Acetate (NH_4OAc) Method

The use of 1 M NH_4OAc was first proposed as a method for CEC determination by Schollenberger and Dreibelbis (1933). Since then it has been reviewed and revised by Schollenberger and Simon (1945), Peech (1945), and Peech et al. (1947). More recently Thomas (1982) listed it as a method for the determination of exchangeable bases.

Reagents

Ammonium acetate (CH$_3$COONH$_{4, cp}$), 1 M, pH 7.0. Weigh 77.1 g of ammonium acetate, NH$_4$OAc, in a 1-L beaker. Add 800 mL distilled water and stir mechanically until all crystals have dissolved completely. If the mixture has been properly prepared, the pH of the solution will be slightly above 7, and only a slight adjustment will be needed. Add dropwise acetic acid and stir constantly to adjust the pH of the solution to pH 7.0. Then transfer the solution quantitatively into a 1-L volumetric flask, and make up the volume with distilled water.

Sodium chloride, (NaCl), pH 2.5. Weigh 10 g of NaCl in a beaker. Add 80 mL distilled water and stir mechanically to dissolve the crystals. Add dropwise HCl and stir constantly to adjust the pH to 2.5. Transfer the solution into a 100-mL volumetric flask, and make up the volume to the mark with distilled water.

Sodium hydroxide, (NaOH), 40%. Dissolve 40 g NaOH in 100 mL of distilled water.

Boric acid (H$_3$BO$_3$), 4%. Dissolve 4 g of boric acid per L of distilled water.

Mixed indicator. Dissolve 0.1 g methyl red and 0.1 g bromocresol green in 250 mL ethanol. Store in an amber-colored flask.

Procedure

Weigh 5.0 g of soil in a 100-mL polyethylene centrifuge tube. Add 25 mL NH$_4$OAc solution, and shake the soil mixture mechanically for 1 h. The supernatant solution is then separated from the soil by centrifugation at 2400 rpm for 30 min. Collect both the clear supernatant and the NH$_4$-saturated soil.

Cation Exchange Capacity Determination

The supernatant is filtered into a 100-mL volumetric flask. The NH_4-saturated soil is washed three times with 20 mL 95% ethanol, by shaking and centrifugation. Each washing is then added to the supernatant in the 100-mL volumetric flask. The extract is made up to volume with distilled water and used for the determination of the exchangeable cations, Ca, Mg, Na, and K. The concentration of the bases is measured by atomic absorption spectrophotometry or by plasma emission spectrometry.

The exchangeable H^+ ion concentration is measured separately by the $BaCl_2$ method discussed in another section of this book.

Calculations

ppm Ca (from AA or ICP) = µg Ca/mL
  = 100 x µg Ca/100mL/5 g soil
  = 20 x 100 x µg Ca/100 g (oven dry) soil

me Ca/100g soil = (20 x 100 x µg Ca)/(1000 x 40/2)

Similarly:

me Mg/100g soil = (20 x 100 x g Mg)/(1000 x 24/2)
me Na/100g soil = (20 x 100 x g Na)/(1000 x 23/1)
me K/100g soil = (20 x 100 x g K)/(1000 x 39/1)

CEC = Σ exchangeable Ca, Mg, Na, K, and H (in me/100g)

$$\% \text{ Base saturation} = \frac{\Sigma \text{ exchangeable Ca, Mg, Na and K}}{\text{CEC}} \times 100\%$$

Amount of Ca/acre furrow slice = me Ca x 20 x 20 lbs/acre
= me Ca x 20 x 20 x 1.12 kg/ha

Amount of Mg/acre furrow slice = me Mg x 12 x 20 lbs/acre
= me Mg x 12 x 20 x 1.12 kg/ha

Amount of Na/acre furrow slice = me Na x 23 x 20 lbs/acre
= me Na x 23 x 20 x 1.12 kg/ha

Amount of K/acre furrow slice = me K x 39 x 20 lbs
= me K x 39 x 20 x 1.12 kg/ha

9.3.2 Alternative Method

The CEC can also be determined by measuring the amount of NH_4^+ adsorbed by the exchange complex. In this case, the NH_4-saturated soil, obtained after washing with ethanol in the above procedure, is mixed with 25 mL NaCl (pH 2.5) solution and shaken mechanically for 30 m. The supernatant is separated from the soil by centrifugation. Using a full pipet, transfer 5 mL of the supernatant into a micro Kjeldahl still. Add 5 mL NaOH (40-50%) and distill the NH_3 into 25 mL of boric acid containing 5-10 drops of the mixed indicator. The boric acid solution will turn green when NH_3 is collected. The solution is then titrated with H_2SO_4 of known normality (N) to the original violet color (= T mL). Do a blank analysis (= B mL).

Instead of using a Kjeldahl still, the NH_4^+ can also be determined by the method using an Auto-Technicon Analyzer, or with the NH_4 specific ion electrode, as discussed in the section of the determination of N.

Calculations

CEC = (T - B) x N = me/5 mL

Cation Exchange Capacity Determination

Complete the calculation for 100 g soil.

Exchangeable H^+ = CEC - exchangeable bases

The concentration of exchangeable bases (in me/100 g) is determined in the first method discussed above.

9.3.3 Determination of ECEC

The effective CEC, or ECEC, is also an important parameter and is used as a criteria for the placement of soils in the U.S. Soil Taxonomy. As defined earlier:

ECEC = (Σme exchangeable bases + me exchangeable Al)/100 g soil

In the U.S. Soil Taxonomy, the value of the ECEC above should be converted into me/100 g clay. The exchangeable bases are defined as the bases displaced by a 1 M NH_4OAc (pH 7.0) solution, whereas the exchangeable Al is the Al extracted by a 1 M KCl solution.

Reagents

The reagents for use in the determination of the exchangeable bases are listed in the NH_4OAc method discussed above.

Potassium chloride, KCl, 1 M, pH 2.0. Weigh 74.6 g of KCl in a 1-L beaker. Add 800 mL distilled water, and stir to dissolve. Adjust the pH to 2.0 by adding dropwise HCl under constant stirring. Transfer the solution into a 1-L volumetric flask, and make up to volume with distilled water.

Procedure
The procedure for the determination of exchangeable bases by the NH_4OAc method is discussed above.

The procedure for the determination of KCl-extractable Al is as follows. Weigh 5 g of soil in a 100-mL polyethylene centrifuge tube. Add 50 mL of the KCl solution, shake mechanically for 3 h, and centrifuge at 2400 rpm for 30 min to separate the supernatant from the soil. Filter the supernatant into a 100-mL volumetric flask. Wash the soil residue thoroughly with distilled water and add the washings to the solution in the volumetric flask. Make up to volume with distilled water, and determine the Al concentration by atomic absorption spectrophotometry or plasma emission spectroscopy.

Calculations
For the calculation of exchangeable bases see the NH_4OAc method.

$$\text{ppm Al (from AA or ICP)} = \mu g\ Al/mL$$
$$= 100 \times \mu g\ Al/100\ mL/5\ g\ soil$$
$$= 20 \times 100 \times \mu g\ Al/100\ g\ \text{(oven dry) soil}$$

$$\text{me Al/100 g soil} = (20 \times 100 \times \mu g\ Al)/(1000 \times 27/3)\ me/100g$$

$$ECEC = (\Sigma\text{me exchangeable bases} + \text{me Al})/100\ g\ soil$$

9.3.4 Determination of CEC_p
As indicated before, the CEC_p is defined as the CEC attributed to the permanent charge of clay minerals. The permanent charge is a negative charge resulting from unsatisfied valences in the crystal structure of clay minerals. It is produced by isomorphous substitution of the Si and Al atoms in the silica-tetrahedrons and Al-octahedrons, respectively. The

Cation Exchange Capacity Determination

method was introduced by Mehlich (1960) for the determination of charge characteristics of soils. A revised version is presented in this section (Mehlich, 1981; Tan and Dowling, 1984).

In order to determine the CEC caused by permanent charges, it is necessary to block the variable exchange sites. This is achieved by acidifying the soils so that the cations on the variable sites are replaced by protons. The exchange reaction can also take place in an acid medium. As stated previously, the exposed OH groups, responsible for variable charges, will not dissociate at low pH, and are relatively non-active.

Reagents

Barium chloride, $BaCl_2$, 0.3 M. Weigh 72.9 g $BaCl_2.2H_2O$ in a 1-L volumetric flask. Add 800 mL distilled water and stir to dissolve. Make up to volume with distilled water.

Calcium chloride, $CaCl_2$, 0.3 M. Weigh 44.1 g $CaCl_2.2H_2O$ in a 1-L volumetric flask. Add 800 mL distilled water and stir to dissolve. Make up to volume with distilled water.

Hydrochloric acid, HCl, 0.1 M. Dilute 6.7 mL concentrated HCl with distilled water to the mark in a 1-L volumetric flask. A stronger HCl solution is required with basic soils or soils containing $CaCO_3$.

Procedure

Weigh 10 g of soil in a 100-mL polyethylene centrifuge tube. Add 30 mL HCl solution, and shake for 15 minutes with a mechanical shaker. Centrifuge, and discard the supernatant. Add to the acidified soil 50 mL of $BaCl_2$ solution. Shake the mixture for 30 min with a mechanical shaker. Centrifuge, and discard the supernatant. Rinse the walls of the centrifuge tube with water, and add 50 mL of distilled water to the Ba-saturated soil. Shake for 5 min with a mechanical shaker, centrifuge, and discard the wash water. Repeat the washing procedure once more, and rinse the walls of the centrifuge tube thoroughly.

Add to the washed Ba-saturated soil 50 mL of the $CaCl_2$ solution. Shake for 30 min with a mechanical shaker. Centrifuge, and filter the supernatant carefully into a 100-mL volumetric flask. Wash the residue by shaking with 30 mL distilled water. Centrifuge, and add the wash water to the solution in the volumetric flask. Make up the volume to the 100 mL mark. Determine the Ba concentration in the extract by atomic absorption spectrophotometry or plasma emission spectroscopy.

Calculations

ppm Ba (from AA or ICP) = µg Ba/mL
= 100 x µg Ba/100 mL/10 g soil
= 10 x 100 x µg Ba/100 g (oven dry) soil

me Ba/100 g soil = (10 x 100 x µg Ba)/(1000 x 137/2) me/100g

CEC_p = Σme exchangeable Ba/100 g soil

9.3.5 Determination of CEC_t

The CEC_t is defined here as the CEC attributed by both the variable and permanent charges. In order to make these exchange sites available for cation exchange, the soil pH is increased to 8.2. The CEC_t, therefore, equals the CEC measured at pH 8.2.

Reagents

Barium chloride, $BaCl_2$, 0.3 *M*. See the procedure for the determination of CEC_p.

Cation Exchange Capacity Determination

Calcium chloride, $CaCl_2$, 0.3 M. See the procedure for the determination of CEC_p.

Barium chloride-triethanolamine, $BaCl_2$ - TEA, 0.2 M.

1. Dissolve 300 g $BaCl_2 \cdot 2H_2O$ in 2 L of distilled water.
2. Dilute 90 mL TEA (s.p. 1.126) in 1 L of distilled water. Adjust the pH to 8.2 by titrating with 1 M HCl. Usually 280 mL HCl is required for this procedure. Make up the volume to 2 L with distilled water.

Mix solutions (1) and (2). If properly prepared, the solution strength is 0.1 M or 0.2 M. To check this, pipet 20 mL of the solution in a 150-mL Erlenmeyer flask. Titrate with the 0.3 M $BaCl_2$ solution. Approximately 40-42 mL $BaCl_2$ is required to reach the end point. If the titration requires more than 40 mL $BaCl_2$, dilute with 0.3 M $BaCl_2$ so that the titre is within 40-42 mL. Protect the $BaCl_2$-TEA solution from CO_2 during storage.

Procedure

Weigh 10 g of soil in a 100-mL polyethylene centrifuge tube. Add 20 mL $BaCl_2$ - TEA, and shake for 15 min with a mechanical shaker. Centrifuge, and discard the supernatant.

Add 30 mL $BaCl_2$ 0.3 M, and shake again for 15 minutes. Rinse the walls of the tube with distilled water, and add 50 mL of distilled water to the Ba-saturated soil. Shake, centrifuge, and discard the wash water. Repeat washing with 50 mL water once more, and rinse the walls of the centrifuge thoroughly.

Add 50 mL $CaCl_2$ 0.3 M, and shake for 15 min. Centrifuge and filter the supernatant solution into a 100-mL volumetric flask. Wash the residue with 30 mL distilled water, centrifuge, and add the wash water to the solution in the volumetric flask. Make up to volume with distilled water.

Determine the Ba concentration in the extract by atomic absorption spectrophotometry or plasma emission spectroscopy.

Calculations

ppm Ba (from AA or ICP) = µg Ba/mL
 = 100 x µg Ba/100 mL/10 g soil
 = 10 x 100 x µg Ba/100 g (oven dry) soil

me Ba/100 g soil = (10 x 100 x µg Ba)/(1000 x 137/2) me/100g

$CEC_t = CEC_{8.2}$ = me exchangeable Ba/100 g soil

$CEC_v = CEC_{8.2} - CEC_p$.

9.4 METHODS FOR DETERMINATION OF EXCHANGEABLE H^+

Two methods will be discussed for the determination of exchangeable H^+, e.g., the $BaCl_2$-TEA method of Mehlich (1960) and the KCl method.

9.4.1 Barium Chloride-TEA Method

Reagents

Barium chloride, $BaCl_2$, 0.3 M. See the procedure for the determination of CEC_p.

Barium chloride-triethanolamine, 0.2 M. See the procedure for the determination of CEC_t.

Hydrochloric acid, HCl, 0.05 M. Dilute 41.5 mL concentrated HCl to 500 mL with distilled water. Pipet 50 mL of this solution into a 1-L volumetric flask, and make up to volume with distilled water. Standardize this acid against 0.05 M NaOH.

Mixed indicator

1. Dissolve 0.1 g bromocresol green in 200 mL 90% ethanol.

Cation Exchange Capacity Determination

2. Dissolve 0.1 g methyl red in 200 mL 90% ethanol.
Mix three volumes of (1) with two volumes of (2).

Procedure

Weigh 10 g of soil in a 100-mL centrifuge tube. Add 20 mL $BaCl_2$-TEA, shake and centrifuge at 2400 rpm for 30 min. Collect the supernatant by filtering the solution into a 150-mL Erlenmeyer flask. Rinse the walls of the tube, and wash the soil by shaking with 25 mL distilled water. Centrifuge, and add the wash water to the solution in the Erlenmeyer. Repeat the washing once more. Rinse the walls of the tube with water, and add both the wash- and rinse-water to the solution.

Add 5-8 drops of mixed indicator, and titrate the solution with the 0.05 M HCl solution to pH = 5.0 or to a pink endpoint (A mL).

Run a blank, using a mixture of 20 mL $BaCl_2$-TEA, 30 mL $BaCl_2$, and 40 mL distilled water (B mL).

Calculations

exchangeable H^+ = (B - A) x 0.05 me/10 g soil
= 10 (B - A) x 0.05 me/ 100 g soil

9.4.2 Potassium Chloride (KCl) Method

Reagents

Potassium chloride, KCl, 1 M. Dissolve 74.6 g KCl in 1 L of distilled water.

Sodium hydroxide, 0.1 M. Weigh 4.0 g NaOH, and dissolve in 1 L of distilled water. Standardized the solution before using it to obtain the proper normality, N ($N = M$).

Phenolphthalein, 1%. Dissolve 1 g phenolphthalein in 100 mL ethanol.

Potassium fluoride, KF, 1 M. Dissolve 58.1 KF in 1 L of distilled water in a plastic beaker. Add 10-15 drops phenolphthalein, and titrate with NaOH to a pink endpoint. The solution complexes Al.

Hydrochloric acid, HCl, 0.1 *M.* See the procedure for the determination of CEC_p.

Procedure

Weigh 10 g of soil in a 100-mL polyethylene centrifuge tube. Add 50 mL KCl solution and shake the tube for 30 min with a mechanical shaker. Centrifuge at 2400 rpm for 15 min, and collect the supernatant by filtering into a 250-mL Erlenmeyer flask. Wash with 50 mL KCl, centrifuge, and add the supernatant to the solution in the Erlenmeyer. Repeat this washing procedure once more.

Add 5-6 drops phenolphthalein, and titrate with NaOH to the first permanent pink endpoint (T mL). Do a blank titration (B mL) for corrections. This step determines the so called KCl-acidity (= exchangeable H^+ + exchangeable Al^{3+}).

Continue the analysis with the titrated solution above by adding 10 mL KF. Titrate with HCl until the pink color disappears. Wait 30 min, and titrate further with HCl to a clear endpoint. This step yields the KCl-extractable Al^{3+}.

Calculations

me exchangeable (H^+ + Al^{3+}) = (T - B) x *N*/10 g soil
= 10 x(T -B) x *N*/100 g soil

me KCl-extractable Al^{3+} = mL HCl x *N*/10 g soil
= 10 x(mL HCl x *N*)/100 g soil

me exch.H^+/100g = me exch.(H^+ + Al^{3+}) - me KCl-extr.(Al^{3+})

Chapter 10

DETERMINATION OF MACROELEMENTS

10.1 DETERMINATION OF SOIL NITROGEN

10.1.1 Forms of Nitrogen in Soils

Nitrogen is an essential nutrient for plant growth. It is taken up by plants in large amounts, whereas its concentration in soils is frequently very small. The Ap horizon of most mineral soils contains on the average 0.15% N, which translates to 3.36 Mg N per ha (3000 lbs/acre). Since soil N is mostly organic in nature, N concentrations in soils increase with increased organic matter contents. For example, the Ap horizon of mollisols, soils characterized by high organic matter contents, may contain as much as 6.72 Mg of combined N per ha (= 6 tons/acre; Stevenson, 1965). Organic soils or histosols are characterized by even higher N content. The N content in peat may amount to 2.5%. However, the N content of the Ap horizons of most cultivated soils is estimated to range between 0.05 - 0.4% N (Bremner, 1965). Because of these low amounts, and because of the tendency for N to be lost easily by decomposition of organic compounds, N is listed as a critical soil constituent by Brady (1990). In contrast, the air above the soil contains 70% N, but this N is unavailable to most plants. Only nitrogen-fixing bacteria can utilize atmospheric N.

Nitrogen exists in soils in two major forms: (1) organic N, and (2) inorganic N. However, for analytical purposes, it is perhaps convenient to distinguish a third form of soil nitrogen: (3) total N. Total N is defined as the sum of organic N and inorganic N. Plants satisfy their nitrogen requirement from the inorganic fraction. The organic fraction serves as a reserve of nitrogen in plant nutrition (Harmsen and Kolenbrander, 1965), and will be released only after decomposition and mineralization of organic matter.

Organic N   Most of the nitrogen in soils is in organic form, and only a small portion is present in inorganic form. More than 90% of all the N in soils are in organic form. Organic N is considered to be closely associated with soil organic matter. Humic and fulvic acid contain 1 to 4% N. Such an association protects the N from rapid decomposition. Under normal conditions, only 2-3% a year of organic N is mineralized and released in soils in inorganic form. The major forms of organic N are amino acids, which make up 20-50% of the organic N. A high percentage of the N in fulvic acids is in amino acid form (Stevenson, 1994). Other N containing compounds in soil organic matter are hexosamine, which make up 5-10% of the organic-N, whereas 1% is in the form of amino sugars, purine, and pyrimidine. Additional organic N compounds, usually detected in very small amounts, include choline, creatinine, allantoin, lignin-NH_4, quinone-NH_4, and quinone-NH_2 (Bremner, 1965). Recently, other types of organic N have been discovered. Stevenson (1994) reported the presence of considerable amounts of acid-insoluble forms and hydrolyzable unknown forms of organic N, which were labeled as *HUN*. Acid-insoluble N is the N-fraction not solubilized by acid hydrolysis. The concentration of acid-insoluble organic N increases with an increase in decomposition and humification of organic matter. Therefore, humic and fulvic acids contain considerable amounts of acid-insoluble N.

Inorganic N   Inorganic N is composed of NH_3-N, NH_4-N, NO_3-N, and NO_2-N. Under normal conditions these types of inorganic-N are continuously formed by decomposition and mineralization of organic-N as conditioned by the *nitrogen cycle*. Most of them are soluble in water, but the concentrations of soluble NH_4-N and NO_3-N are estimated to be only 1-2% of the total N in soils. However, their concentrations may vary considerably depending on several conditions, including the application of N-fertilizers. The inorganic N content in bare soils can be low as low as 10 mg/L in winter when no fertilizers have been applied, and reaches values of 40-60 mg/L during spring and summer time, because of increased biological activities. In soils rich in organic matter, even higher values have been reported, such as in peat soils where the inorganic-N content can be as high as 500 mg/L (Harmsen and Kolenbrander, 1965).

NH_4-N is also present in soils as *fixed NH_4*. NH_4^+ ions can be trapped in the intermicellar spaces of expanding 2:1 clay minerals, e.g., smectite, and clay minerals with high intermicellar charges, such as vermiculite. This process, called *ammonium fixation*, can fix 10-40% of the total N in soils. Evidence has been presented that a dynamic equilibrium exists between fixed, exchangeable, and free NH_4^+ ions (Nommik, 1965). Fixation of NH_4^+ occurs, when the equilibrium condition is disrupted by increasing the free NH_4^+ ion concentration, which in turn increases the amount of exchangeable NH_4^+. On the other hand, it is also possible to release all the fixed NH_4^+ by removing the free NH_4^+, such as by uptake by plant roots, or by leaching.

NH_3-N is present in the gaseous form and as *fixed NH_3*. Although the chemical reaction of NH_3 and soil organic matter is also called *NH_3 fixation*, its reaction is completely different from *NH₄ fixation* by clay (Stevenson, 1986). The reaction of lignin and other types of organic compounds with NH_3, known already for some time, has found practical application today in the conversion of organic residue and coal products into N-fertilizers by treatment with NH_3. Application of anhydrous NH_3 fertilizer can also result in fixation of considerable amounts of NH_3 by soil constituents.

Total N As indicated above, total N is composed of organic N and inorganic N, and its content in soils is usually very small. Total N content may range from 0.05% to 0.4% in the Ap horizon of most soils, although mollisols and histosols are characterized by higher N content, as indicated earlier.

Because of the above, the following methods of determination of soil N will be discussed: (1) determination of total N, (2) determination of exchangeable NH_4, (3) determination of NO_3, and (4) determination of NO_2.

10.1.2 Determination of Total N

The determination of total N is a complicated process, because of the presence of various forms of N, and because of the low concentrations of N present in soils. Several methods are available for the determination of total N in soils: (1) Kjeldahl method, (2) Dumas method, (3) NIRS (Near Infrared Reflectance Spectroscopy) method, and (4) Direct Distillation method (Bremner and Mulvaney, 1982; Jones, Jr., 1991). The Dumas method is a dry combustion method, and is the traditional and oldest method for measuring total-N in soils (Dumas, 1831). However, of all the methods used, the Kjeldahl method, which is a wet oxidation method, is perhaps the most common method for determination of total N. The advantages of the Kjeldahl method, (1) the ease of running multiple analyses, and (2) applicability to samples low in N, mark it as a

Determination of Soil Nitrogen

method of first importance. Therefore, this method will be discussed in this book in detail.

10.1.3 Kjeldahl Method

The Kjeldahl method was reported by Johan Kjeldahl in 1883, two years after Dumas (1831) has introduced his dry combustion method. In outline, the Kjeldahl method consists of three phases: (1) the digestion phase, (2) the distillation of NH_4, and (3) the determination of NH_4. In the digestion phase, the sample is decomposed by H_2SO_4, with Na_2SO_4 and a suitable catalyst. The N and C in the sample are converted into $(NH_4)_2SO_4$ and CO_2, respectively. In the distillation phase, the NH_4^+ is converted to NH_3 and distilled into H_3BO_3 or a standard HCl solution. The determination of NH_4^+ is usually done by titration.

Accuracy and precision of Kjeldahl analysis depend on the digestion phase, since a number of organic compounds can not be broken down completely by the procedure. Several organic compounds are refractory to Kjeldahl digestion. Conspicuous among these substances are complex heterocyclic N substances, such as pyridine, nicotine acid, quinoline, certain alkaloids, and several amino acids, e.g., lysine, histidine, tyrosine, and tryptophane. Azo-, diazo-, nitro-, and nitroso-compounds, hydrazines, hydrazones, and oximes are additional compounds that are refractory to Kjeldahl digestion. Fortunately, in many soil and plant samples these compounds are either absent or present in minor amounts, and if present are accompanied by other organic compounds that assist in their decomposition by Kjeldahl digestion.

Because of the difficulties in breaking down several of the organic compounds, it is the digestion procedure that has undergone many revisions since the introduction of the Kjeldahl method. During the years, the method of analysis has also been changed from a macro into a semi-micromethod. The usual 800-mL Kjeldahl flasks heated by a gas flame have been replaced by

50-mL electrically heated flasks. Today *block digestors*, which can accommodate a large number of 50-mL tubes, are preferred. Block digestion systems are commercially available in various sizes, and require only a small space in the laboratory. This semimicro Kjeldahl method requires smaller amounts of reagents, such as smaller amounts of concentrated H_2SO_4, heavy metals used in the catalyst, and NaOH, which make disposal of these reagents after use more convenient.

Sample Size and Preparation

The size of sample required for Kjeldahl analysis depends on its N content. The following amounts are usually satisfactory for macro-Kjeldahl procedures:

Material	Weight in g
Legume grains	0.5
Legume hay	1
Cereal grains	1
Cereal straw	5
Peat and muck	2-8
Manures	1-4
Fertilizers	0.7-3.5
Soil	5-10

In semimicroanalysis, the sample size is 1/10 that of a macro sample. However, it must be realized that the larger the amount of samples used, the higher the precision and accuracy of analysis. To achieve the proper homogeneity with the use of small amounts of samples, soil samples have to be ground to pass a 0.4-mm (40-mesh) sieve. Plant samples are usually ground through a Wiley mill to pass a 0.85-mm (20-mesh) sieve.

Determination of Soil Nitrogen

Digestion Phase

A number of reagents are used in the digestion phase, such as salicylic acid, Zn dust, K_2SO_4 or Na_2SO_4, a suitable catalyst, and concentrated H_2SO_4.

1. *Salicylic Acid or Zn Dust*

For total N determination special provision must be made to include NO_3-N and NO_2-N. NO_3-N will not be recovered in the Kjeldahl digestion process, whereas NO_2^- will decompose by the addition of concentrated H_2SO_4 and is lost in the air. In many instances, however, the NO_3^- and NO_2^- content in soils are so small that to all intents and purposes they can be ignored, and the value for organic N and NH_4-N taken as total N.

When NO_3-N must be included, the sample is pretreated by soaking in the cold, preferably overnight, in concentrated H_2SO_4 in which salicylic acid has been dissolved. The resulting nitro-salicylic acid is reduced by heating gently with $Na_2S_2O_3$. Zn dust may also be used for reduction of nitro-salicylic acid. Considerable care and attention are required during the reduction process with $Na_2S_2O_3$ because the mixture tends to froth badly and some losses of N may occur accordingly. Where the addition of salicylic acid and reducing agent can be avoided, a considerable saving of time is affected. It should be noted that this method of pretreatment for including NO_3-N cannot be used on wet samples.

2. *Potassium Sulfate or Sodium Sulfate*

K_2SO_4 or Na_2SO_4 is added to increase the boiling temperature of H_2SO_4 (330°C) to 360-390°C. In addition, the SO_4^{2-} anion is needed for preventing loss of NH_3 from the mixture, by conserving it as $(NH_4)_2SO_4$. The use of K_2SO_4 is preferred, since Na_2SO_4 tends to result in spattering, although Na_2SO_4 will yield a higher boiling temperature than K_2SO_4. According to Bremner (1960) the temperature is very important in the digestion process. Loss of N will likely occur when the digestion temperature exceeds 400°C. The digestion temperature is controlled largely by the amount of K_2SO_4 used. With the addition of low amounts of K_2SO_4 (0.3 g/mL of H_2SO_4), a longer digestion time is needed to reach the temperature for obtaining accurate results. In the presence of higher amounts of K_2SO_4 (1.0 g/mL of H_2SO_4), shorter digestion periods will produce satisfactory results. More shorter digestion time using high amounts of K_2SO_4 is considered of more advantage over longer digestion time. Bumping, spattering, and other problems may arise from using longer digestion periods. However, the use of high amounts of K_2SO_4 may create frothing during digestion, and may tend to solidify the digest upon cooling. When solidification occurs after cooling, more time is perhaps required to take up the digest with water for further analysis. Bremner and Mulvaney (1982) reported that solidification did not necessarily result in a loss of N.

3. *Catalyst*

Originally, $KMnO_4$ (potassium permangate) was used by Johan Kjeldahl as a catalyst to speed up the digestion process. However, during the years this has been replaced by a number of other catalysts. Mercury, HgO, HgI_2, Se, $SeOCl_2$, $CuSO_4$, V_2O_5, or Fe is found to be more effective as a catalyst than is $KMnO_4$. The addition of phosphate, H_2O_2, $K_2S_2O_8$, or $HClO_4$ as aids in digestion has also been proposed. The catalyst most commonly employed is Hg, HgO, Se or $SeOCl_2$, or mixtures of these metals or compounds. Mercury, Hg, and selenium, Se, are considered to be the most effective catalysts, and the combination of Se, $CuSO_4$ and HgO appears to have the best catalytic action. However, both Hg and Se are very toxic metals, and pose serious problems for their safe disposal in the environment. Titanium oxide, TiO_2, is suggested as a substitute for HgO. When used in combination with $CuSO_4$, it is reported that digestion time for animal feeds is only 40 minutes (Jones, et al., 1991). Today, H_2O_2 is frequently used for digestion of

Determination of Macroelements

plant samples, and the use of catalysts containing hazardous metals is hereby completely eliminated.

Recently, commercially prepared mixtures are available in the form of pellets (*Kjeltabs*), tablets (*Kelmate*), or digestible bags. Although these mixtures still contain Hg and Se, they are very convenient for use in Kjeldahl digestion. For the chemical composition of some of these commercially prepared catalysts, reference is made to Jones et al. (1991).

4. *Sulfuric Acid*

Sulfuric acid, H_2SO_4, is the essential reagent for the digestion process. It is consumed during the oxidation of organic matter, and during the reaction with mineral soil constituents. Excessive losses of H_2SO_4 should be avoided since this may result in the decomposition of $(NH_4)_2SO_4$ with a consequent loss of NH_3. More H_2SO_4 may be added at any time as becomes necessary. Excessive losses of H_2SO_4 may arise from (1) excessive heat, which results in unnecessary volatilization of the H_2SO_4, and (2) underestimation of the consumption of H_2SO_4 by either the organic matter or K_2SO_4. However, with proper regards to principles and techniques, excessive losses of H_2SO_4 due to volatilization and decomposition can be prevented.

The amount of H_2SO_4 needed for the oxidation varies with the type of sample analyzed. Estimates of the quantities of H_2SO_4 consumed by various materials during the digestion process are listed below (Bremner and Mulvaney, 1982):

Sample	H_2SO_4 (36 N)
1 g carbohydrates	4.0 mL
1 g protein	5.0 mL
1 g fat	10.0 mL
1 g K_2SO_4	0.5 mL
1 g salicylic acid	6.8 mL
1 g $Na_2S_2O_3 \cdot 5H_2O$	0.5 mL
1 g soil organic matter	5.8 mL

From this list, it is obvious that more H_2SO_4 will be required for the digestion of samples high in fat.

The loss of N on continued digestion is perhaps a general phenomenon in Kjeldahl digestion. These losses seem to be more pronounced when Se is used, and to be somewhat reduced when HgO accompanies Se. The only means for reducing such losses is to use adequate amounts of H_2SO_4.

5. *Digestion Time*

Digestion occurs in a two-phase sequence. In the first phase of digestion, the mixture turns black in color, and large amounts of fumes are produced. During the second phase, the mixture first turns light brown in color, to become almost colorless at the end. However, complete digestion is not coincident with *clearing time*, since as much as 10% of the N in the samples has yet to be converted into NH_4. For complete digestion, the process of digestion is allowed to continue after clearing time has been reached. Usually the total time required is taken as 1.5 times the clearing time. If Cu is used in the catalyst mixture, the digest has a light blue-green color, mixed with white precipitates, after cooling.

Determination of Ammonium

The determination of ammonium in the digest is traditionally done by distillation and titration. NH_3 is removed from the digest by distillation from alkaline solution:

$$(NH_4)_2SO_4 + 2NaOH \rightarrow 2NH_3 + Na_2SO_4 + 2H_2O$$

The NH_3, collected in a saturated solution of H_3BO_3, is determined by titration with a standard solution of H_2SO_4 or HCl. However, with the recent advancement in technology the ammonium in the digest is more conveniently determined by spectrophotometry or colorimetry. The use of the ammonium specific-ion electrode has attracted attention for a while, but has found limited applications. Of the three methods stated above, the most popular method today is the colorimetric method. Two major colorimetric methods are available: (1) the *phenol-hypochlorite method*, and (2) *salicylate-hypochlorite method*. The basic principle in the colorimetric method is, that in alkaline solution, NH_3 forms a blue color with a phenol-sodium hypochlorite reagent. The intensity of the blue color is proportional to the amount of NH_3 present. This analytical technique, known later as the *indophenol blue* or *phenate method*, forms the basis for the rapid analysis of NH_3 with automated colorimetric instruments, e.g., Technicon AutoAnalyzer (Isaac and Johnson, 1976), or Flow-injection Analyzer (Smith and Scott, 1990; Jones et al., 1991). The color intensity is measured by a colorimeter, and the results printed by a recorder. Routine analysis by the traditional Kjeldahl digestion-distillation method requires the close attention of a skilled technician. With the development of these automated instruments, the Kjeldahl analysis can be performed just as effectively and with just as much precision and accuracy by far less skilled personnel. The automated procedures have gained considerable popularity because of their speed, sensitivity, and ease of use. They are also frequently combined for determination of nitrite, nitrate, and phosphorus.

Since phenol is known today as a harmful carcinogenic compound, this reagent tends to be replaced by less toxic compounds, such as in the *salicylate-hypochlorite method*. The color reaction is based on the reaction of NH_4 with a weakly alkaline mixture of sodium salicylate ($NaC_7H_5O_3$) and chlorine (Cl), supplied in the form of Clorox. Sodiumnitroprusside $[Na_2Fe(CN)_5NO \cdot 2H_2O]$ is added to speed up the color reaction at room temperature. When a Technicon AutoAnalyzer is used, a wetting agent (Brij-35) is added to the buffer solution and the sodium salicylate mixture.

1. *Alkaline Distillation and Titration of Ammonium*

Reagents
1. Concentrated H_2SO_4.
2. Standardized 0.01 N H_2SO_4.
3. Kjeltab, Kjelmate, or make your own catalyst mixture by mixing 200 g K_2SO_4, 20 g $CuSO_4 \cdot 5H_2O$, and 2 g Se.
4. NaOH 10 N.
5. Saturated boric acid, H_3BO_3, solution (50 g/L).
6. Mixed indicator. Dissolve 100 mg of methylene blue and 66 mg methyl red in 100 mL ethanol. This indicator is green in alkaline solution, gray at the neutral point, and purple in acid solution.

Procedure

Weigh 1.000 g of soil, ground to pass a 0.4-mm sieve, in a filter paper. More sample can be weighed for soils low in N. Fold the filter paper into a small bag, and drop it in a digestion block tube. Add Kjeltab, Kjelmate, or 1.5 g catalyst mixture, and 5 mL concentrated H_2SO_4. Place the tube in a digestion block, and heat to boiling until green. Continue heating for another hour. After digestion is completed, allow the tube to cool, and slowly add 20 mL distilled water. The digest is ready for NH_4 determination by conventional distillation-titration procedure, or by automated colorimetric methods.

In case of plant samples, weigh 250 mg of sample, ground with a Wiley mill, add 10 mL H_2SO_4, and proceed as discussed above. (Frequently, the catalyst mixture can be replaced and digestion is performed by adding 2 mL 30% H_2O_2.) Continue heating for 30 more minutes, until white fumes are produced. Then cool the tube, and repeat digestion with 1 mL H_2O_2 until the digest is clear and remains clear on cooling. Dilute with 15 mL distilled water, and the digest is ready for NH_4 determination.

Distillation and Titration Process

Place a 50-mL Erlenmeyer flask containing 10 mL H_3BO_3 and a few drops of mixed indicator under the condenser stem of the distillation apparatus, so that the end of the stem is touching the surface of the H_3BO_3 solution. Then, transfer and wash the entire content of the digestion tube into the steam chamber of the distillation apparatus. Add slowly 20 mL of 10 N NaOH into the digest. When about 1 ml of NaOH remains in the funnel stem, rinse the funnel quickly by spraying with water, and close the chamber rapidly. Start distillation by allowing steam to flow through the digest, and as soon as NH_3 starts to collect, the boric acid solution turns green. When the distillate reaches the 35 mL mark on the receiver Erlenmeyer flask, stop the distillation. Rinse the end of the condenser stem, and titrate the distillate with standardized 0.01 N H_2SO_4, dispensed from a 10-mL microburette. Run also a blank analysis. The total N content in percentages is calculated as follows:

$$\% N = [(T - B) \times N \times 1.4]/s$$

in which: T = sample titration, mL standard acid, B = blank titration, mL standard acid, N = normality of standard acid, and s = sample weight, g.

2. *Colorimetric Determination of Ammonium*

In the absence of automated instruments, the NH_4^+ in the digest can also be measured manually with a colorimeter. This manual procedure is in fact the basic principle for use in the automated instruments.

Reagents
1. Sodium salicylate-Sodium nitroprusside solution: Dissolve 300 mg of Na-nitroprusside [$Na_2Fe(CN)_5NO \cdot 2H_2O$] and 150 g Na-salicylate ($NaC_7H_5O_3$) in 600 mL water.

2. Sodium hypochlorite (NaOCl, or Clorox): Measure 6 mL Clorox, containing 5.25% Cl, in a 100-mL volumetric flask, and dilute to the mark with distilled water.
3. Buffer solution: Dissolve 50g of Na-K-tartrate ($NaKC_4H_4O_6 \cdot H_2O$) and 26.8 g disodium phosphate (Na_2HPO_4) in 600 mL distilled water in a 1-L volumetric flask. Add 54 g NaOH, and allow this to dissolve by constant stirring, before the volume is made up to 1 L with distilled water.
4. Standard NH_4^+ solution: Dissolve 412.5 g of ammonium sulfate $[(NH_4)_2SO_4]$ in 1 L of distilled water. This solution contains 100 µg of NH_4^+/mL. Pipet 5 mL of this solution in a 250-mL volumetric flask, and dilute to the mark with distilled water. The resulting solution, containing 2 µg NH_4 per mL, is used for making a standard (calibration) curve.

Procedure

Pipet 1 mL of the diluted Kjeldahl digest into a colorimetric tube or cuvette. Add 5.5 mL buffer solution, and 4 mL Na-salicylate-nitroprusside solution and stir to mix. Add 2 mL Na-hypochlorite solution and mix. Allow the solution to stand for 45 minutes at 25 C for complete color development, and measure the absorbance of the colored solution at 650 nm.

Prepare a standard curve with 4 or 5 different known amounts of NH_4^+ using the same procedure as outlined above.

10.1.4 Determination of Exchangeable NH_4^+

Reagent

KCl solution, 2 M: Dissolve 149 g KCl in 800 mL distilled water in a 1-L volumetric flask, and dilute to the mark with distilled water.

Determination of Soil Nitrogen

Extraction of Exchangeable NH_4^+

Weigh 5 g of soil in a centrifuge tube, and add 50 mL of 2 *M* KCl. Stopper the tube, and shake it on a mechanical shaker for 1 hr. Centrifuge the suspension to separate the soil from the supernatant liquid, and filter the latter into a polyethylene vial for storage and further analysis. If the extract cannot be analyzed immediately, store the filtrate in a refrigerator.

Procedure

Pipet 1 mL of the KCl extract into a colorimetric tube and proceed as described above for the determination of total N. Some scientists suggest adding 1 mL EDTA solution to the KCl extract before the addition of the sodium salicylate-hypochlorite buffer solution (Keeney and Nelson, 1982). EDTA is used with the purpose to chelate the metals so that they will not affect the color reaction.

10.1.5 Determination of NO_3^-

The accurate determination of NO_3^--N in soils requires that considerable care be taken in handling the samples between the field and the laboratory. Unless special provisions are made, NO_2^--N is usually included as NO_3^--N. The determination of NO_3^- can be performed by *indirect* and *direct* methods:

Indirect Methods Through Conversion of NO_3^- to NH_4^+

1. Distillation with and without MgO or Devarda alloy. The distillation procedure as described above for the determination of total N is performed with and without the addition of Devarda alloy. The difference in results equals the NO_3^- concentration in the sample.
2. Cu-Cd reductor method. NO_3^- is reduced into NO_2^- by passing the extract through a Cu-Cd reductor column. The NO_2^- is then determined by the *Griess-Ilosvay* method employing manual or automated procedures.

Direct Methods
1. Colorimetric method using the phenol disulfonic acid procedure. This method is based on the development of a yellow color, when 1,2,4 phenol disulfonic acid is nitrated and the resulting 6-nitro-1,2,4 phenol disulfonic acid is neutralized with NH_4OH. The intensity of the yellow color is proportional to the NO_3^- concentration in the sample.
2. Nitrate specific-ion electrode.

Extraction of NO_3^-
Since nitrate, NO_3^-, is soluble, all reagents are suitable for extraction of nitrate from soil, even water. Acidified 1 N KCl, pH 1.0, and neutral 1 N KCl or NaCl were used (Bremner, 1965). Keeney and Nelson (1982) proposed the use of 2 M KCl, as described above for the extraction of exchangeable NH_4^+. Jackson (1958) used 1 N $CuSO_4$ solutions, whereas Bremner used 500 mg $CaSO_4$ dissolved in 250 mL water for extraction of NO_3^-.

Procedures

Phenol Disulfonic Acid Procedure

Reagents
1. Phenol disulfonic acid solution: Dissolve 25 g of pure phenol (crystals) in 150 mL concentrated H_2SO_4 in a 500-mL beaker. Add 75 mL fuming H_2SO_4, mix the solution, and place the beaker in boiling water for 2 hr. Store the resulting phenol disulfonic acid, $C_6H_3OH(HSO_3)_2$ in an amber colored bottle. This reagent is highly corrosive.

Determination of Soil Nitrogen

2. Powdered $CaCO_3$, reagent grade.
3. Powdered $CaSO_4 \cdot 2H_2O$, reagent grade.
4. Ag_2SO_4, for removal of Cl^-, which interferes with the color reaction.
5. Powdered $Ca(OH)_2$, reagent grade.
6. Powdered $MgCO_3$, reagent grade.
7. Standard NO_3^- solution: Weigh 721.4 mg of pure dry KNO_3 in a 1-L volumetric flask, and dissolve with distilled water to the mark. This solution, containing 100 mg N/L, is called the stock solution. Prepare a working solution by pipetting 25 mL of the stock solution into a 250-mL volumetric flask. This working solution contains 10 mg N/L, and is used for making a standard curve.

Preparation of Cl⁻-free Extract

Weigh 50.0 g of soil (25.0 g of histosol) in a 500 mL Erlenmeyer flask, and add 500 mg $CaSO_4$ and 250 mL distilled water. Shake the flask mechanically for 15 min, and centrifuge the content to separate the supernatant from the soil. Filter the clear supernatant liquid into a clean 500-mL Erlenmeyer flask, add 200 mg $AgSO_4$, and shake the flask mechanically for 15 min. Add 200 mg $Ca(OH)_2$ and 500 mg $MgCO_3$ to precipitate Ag^+. After shaking the mixture for 5 min, filter the suspension into a polyethylene vial for further analysis.

Analysis of the Extract

Pipet 20 mL of the clear extract into a 150-mL Pyrex beaker, and evaporate the extract to dryness. Allow the beaker to cool to room temperature, and add rapidly 3 mL of phenol disulfonic acid. Rotate the beaker to effect complete contact, and allow the color reaction to develop (10 min). Add 20 mL distilled water and stir to dissolve. Transfer the colored solution into a 100-mL volumetric flask, dilute to the mark with distilled water, stir thoroughly, and measure the color intensity at 420 mµ against a blank solution.

Prepare a standard curve by pipetting different aliquots of the working solution and proceeding according to the procedure as described above.

10.1.6 Determination of NO_2^-

Nitrite can also be determined by using indirect and direct methods.

Indirect Procedure: distillation and titration method.
1. Determine $(NH_4^+ + NO_3^- + NO_2^-)$ by adding Devarda alloy and MgO in the distillation process.
2. Determine $(NH_4^+ + NO_3^-)$ by adding sulfamic acid before the addition of Devarda alloy and MgO to destroy NO_2^-.
3. The difference between analyses (1) and (2) is the NO_2^- concentration.

Direct Procedure: Griess-Ilosvay method (Bremner, 1965; Keeney and Nelson, 1982).

Procedure

Modified Griess-Ilosvay Method

Reagents
1. Diazotizing solution: Weigh 500.0 mg of sulfanilamide in a 100-mL volumetric flask. Add 80 mL 2.4 N HCl solution to dissolve the reagent under constant stirring, after which the volume is made up to the mark with the HCl solution. Store this solution at 4°C in a refrigerator.
2. Coupling agent: Weigh 300.0 mg N-(1-naphthyl)-ethylenediamine hydrochloride in a 100-mL volumetric flask. Add 80 mL of 0.12 N HCl and dissolve the reagent under constant stirring, after which the solution is made up to volume with 0.12 N HCl. Store the solution in an amber colored bottle in

Determination of Macroelements

a refrigerator at 4°C.

3. Standard NO_2^- solution: Weigh 246.4 mg $NaNO_2$ in a 1-L volumetric flask and add 800 mL distilled water to dissolve, after which the volume is made up to the mark. This solution, containing 50 mg N/L, is the stock solution and must be stored in a refrigerator at 4°C. To prepare a working solution for a standard curve, pipet 20 mL of the stock solution into a 1-L volumetric flask. The volume is made up to the mark with distilled water.

Analysis of the Extract

The 2 M KCl extract (see 10.1.3) can be used for the determination of NO_2^-. The extract is sometimes colored, but this does not interfere with the analysis (Keeney and Nelson, 1982).

Pipet 2 mL of the 2 M KCl extract into a 50-mL volumetric flask, and add 40 mL of distilled water. Add 1 mL of diazotizing reagent, mix by swirling the flask, and allow the mixture to react for 5 min, after which 1 mL of coupling agent is added. Mix the solution and allow it to stand for 20 min. Make up to volume with distilled water, mix thoroughly, and measure the color intensity at 540 nm against a blank solution.

Prepare a standard curve, by pipetting 0, 1, 2, 4, and 5 mL of the working solution into 50-mL volumetric flasks, and proceed as described above.

10.2 DETERMINATION OF SOIL PHOSPHORUS

10.2.1 Forms of Phosphorus in Soils

Phosphorus is an essential plant nutrient, and is taken up by plants in the form of the inorganic ions: $H_2PO_4^-$, and HPO_4^{2-}. It is needed for root development, for providing vigor in straw of cereal crops, for seed formation, and for controlling plant maturity. Phosphorus is also an essential component of *ADP* (adenosine diphosphate) and *ATP* (adenosine triphosphate), compounds playing a vital role in photosynthesis and ion uptake and transport in plants. Phosphorus is also an essential constituent of nucleic acids (DNA and RNA). These acids control the synthesis of enzymes and proteins and are also responsible for the genetic transfer in cell division.

The total P content in soils varies considerably from soil to soil and from region to region. Soils from acid rocks are usually low in P, whereas soils from basic rocks are high in P. Mollisols and alfisols are reported to possess high to moderate amounts of P, whereas ultisols are usually considered to be low in P. The soils in the southern coastal plain of the U.S. are generally low in total P, and contain 0.02 to 0.04% P (equivalent to 20 - 40 mg P/100g). On the other hand, the soils in the northwestern region of the U.S. are reported to be high in P, with a P content ranging from 0.09 to 0.13%. On the average, the total P content in surface soils ranges from 50 to 80 mg P/100g (Stevenson, 1986). A P content of 50 mg/100g is equivalent to 1120 kg P/ha (1000 lbs/acre).

Phosphorus occurs in soils in inorganic and organic form, and both forms will affect the amount available to plants. However, availability of soil P depends on a large number of factors, e.g. dissolution into ionic forms, soil pH, decomposition of organic matter, and microbial activity. A considerable amount of time and effort have been spent to develop chemical procedures for extracting the available fraction of soil P. Therefore, from the standpoint of soil analysis and plant nutrition, it seems necessary to distinguish also *available P* as a third category of soil P.

The general perception in the past is that most of the soil P is in organic form. This is not always true, as can be noticed from Table 10.1. Tisdale and Nelson (1993) reported that in general the inorganic P content in soils is higher than the organic P content. In solution, phosphorus is present only in the inorganic form, as either the primary, $H_2PO_4^-$, or the secondary,

HPO_4^{2-}, orthophosphate ion. In acid soils, $H_2PO_4^-$ will be dominant over HPO_4^{2-}, whereas at pH>7, HPO_4^{2-} will be dominant together with some PO_4^{3-}. The concentration of these ions in the soil solution depends on the pH, and is estimated to be 1 mg/L

Table 10.1 Total P and Inorganic and Organic P Fractions in Selected Soils.

Soils	Total P mg/kg	Inorganic P%	Organic P%
Alfisols	0524	45	55
Andisols	4700	63	37
Aridisols	0703	64	36
Mollisols	0559	41	59
Oxisols	1414	81	19
Spodosols	0398	35	65
Vertisols	0362	14	86

Source: Brady (1990); Soltanpour et al. (1988).

or less. Maximum availability of these phosphate ions for plant growth occurs within the pH range of 5.5-6.5 (Tisdale and Nelson, 1993).

Inorganic P Fraction

The inorganic P fraction in soils includes a very large group of minerals. Most of their members are so rare that they are seldom mentioned. In general, two major groups of inorganic phosphate compounds can be recognized in soils: (1) the variscite-strengite group, composed of aluminum phosphate (variscite), iron phosphate (strengite), and their intergrades, and (2) the *apatite group*, composed of calcium phosphate minerals. The variscite-strengite group is usually present in acid soils, where conditions are favorable for the presence of Al and Fe. On the other hand, the apatite group is usually formed in neutral to basic soils, where conditions are favorable for the presence of Ca. Three apatite minerals can be distinguished: *fluorapatite, chlorapatite,* and *hydroxylapatite*, containing F, Cl, and OH, respectively. Of the three, fluorapatite is the most common. The apatite mineral is used in the manufacture of phosphate fertilizers. To render the P more soluble, the mineral is treated with sulfuric acid, H_2SO_4, and converted into the fertilizer *superphosphate*. The residue from the production of superphosphate contains gypsum, $CaSO_4$, a valuable source of liming material.

Organic P Fraction

As indicated earlier, the concentration of organic P is very variable, and may range from zero to over 2% P (Anderson, 1980). The organic P content of soils increases with increased organic matter content. A positive linear regression has been detected between organic P and organic C content (Stevenson, 1986).

The organic P fraction is divided generally into 4 major groups: (1) *inositol phosphates*, a series of phosphate esters formed from the carbohydrate inositol, $C_6H_{12}O_6$, and P compounds, (2) *nucleic acids*, RNA and DNA, components of animal and plant cell nucleus, (3) phospholipids, derivatives of glycerol occurring in all living cells, which are soluble in *fat* solvent, and (4) miscellaneous esters, making up the remainder of the organic P in soils. The inositol phosphate fraction, making up 10-50% of the total organic P content, is the dominant fraction. Phytic acid, an inositol phosphate occurring widely in nature, is very stable in alkaline conditions, but will hydrolyze in acid medium (pH 4.0). Phospholipids and nucleic acids account for only 1-5% and 0.2-2.5%, respectively, of the total organic P content, whereas the other esters are present only in trace amounts.

Determination of Macroelements

Available P

As discussed above the total P content does not reflect the amount of P in soils available to plants, and numerous efforts have been made to define that fraction of P in soils that is available for plant growth. This fraction, called *available P*, can be defined in several ways depending on the method of extraction. Several of the established extractants in the past are listed in Table 10.2 (Kamprath and Watson, 1980; Jones, 1985). Today seven types of available P have been formulated according to the more commonly used extraction procedures (Olsen and Sommers, 1982; Stevenson, 1986).

Table 10.2 Reagents Used for Extraction of Available P

Extracting reagents	Soil/reagent ratio	Name of procedure
$0.025N$ HCl + $0.03N$ NH$_4$F	1:10	Bray 1
$0.1N$ HCl + $0.03N$ NH$_4$F	1:17	Bray 2
$0.5M$ NaHCO$_3$, pH 8.5	1:20	Olsen
$0.05N$ HCl + 0.025 H$_2$SO$_4$	1:4	Mehlich 1
$0.2N$ CH$_3$COOH + $0.2N$ NH$_4$Cl + $0.015N$ NH$_4$F + $0.012N$ HCl	1:10	Mehlich 2
$0.2N$ CH$_3$COOH + $0.25N$ NH$_4$NO$_3$ + $0.015N$ NH$_4$F + $0.013N$ HNO$_3$ + $0.001M$ EDTA	1:10	Mehlich 3
$0.002N$ H$_2$SO$_4$ buffered at pH 3 with (NH$_4$)$_2$SO$_4$	1:100	Truog
$0.54N$ HOAc + 0.7 NaOAc, pH 4.8	1:10	Morgan
$0.02N$ Ca-lactate + $0.02N$ HCl	1:20	Egner
1% citric acid	1:10	Citric acid

1. *Phosphorus Soluble in Water*
The P in soil is extracted with water, since a water extract closely approximates the soil solution. Phosphate ions soluble in water are expected to be easily available for plant uptake.

2. *Phosphorus Soluble in Dilute HCl+NH₄F*
With this procedure, also known as the *Bray 1 method*, easily acid-soluble P, originating from Ca-phosphates and Al- and Fe-phosphates, is extracted from soils. NH_4F chelates Al and Fe, and dissolves in this way the phosphate mineral. This extraction is successful for acid soils.

3. *Phosphorus Soluble in NaHCO₃*
This fraction of P is extracted by 0.5 M $NaHCO_3$, pH 8.5. The procedure is also known as the *Olsen method*, and is used mostly for extraction of available P from calcareous or alkaline to neutral soils. Due to the high pH, this reagent also dissolves part of the organic P. Therefore, the amount of P extracted also includes organic P or *labile P*.

4. *Phosphorus Soluble in Dilute HCl+H₂SO₄*
This fraction of P is extracted with a mixture of 0.05 N HCl and 0.025 N H_2SO_4. The procedure, called the *double acid procedure* or *Mehlich No.1 procedure*, is developed for use with acid soils of the southern Piedmont. The method is also useful for extraction of major cations from acid soils (Mehlich, 1984).

5. *Phosphorus Soluble in NH₄HCO₃+DTPA*
This is available P, extracted by a mixture of 1 M NH_4HCO_3 and 0.005 M DTPA (diethylenetriaminepentaacetic acid) at pH 7.6. DTPA chelates and dissolves the micronutrients at the same time, hence the method is also applicable for determination of Fe, Cu, Mn, and Zn.

Determination of Macroelements

6. *Anion Resin-Extractable P*

Phosphate ions are negatively charged and will be adsorbed by anion resins without chemical or pH changes. The amount of P desorbed is considered available for plant growth.

7. *Phosphorus Determined by the Isotopic Dilution of ^{32}P*

The hypothesis is that P in the solid-phase is in equilibrium with P in the solution-phase:

$$P_{solid} \rightleftharpoons P_{solution}$$

Both these native P's are in the form of ^{31}P, which is sometimes called *labile* P. Therefore labile P = sum of P_{solid} and $P_{solution}$. Addition of ^{32}P to the soil forces the equilibrium to change as follows:

$$^{31}P_{solid} + {^{32}P_{solution}} \rightleftharpoons {^{32}P_{solid}} + {^{31}P_{solution}}$$

At equilibrium the ratio of the two isotopes in the solid phase equals the ratio of the isotopes in solution, hence the equilibrium constant = 1. The P_{solid} or $^{31}P_{solid}$ can then be calculated as follows:

$$^{31}P_{solid} = ({^{32}P_{solid}}/{^{32}P_{solution}}) \times {^{31}P_{solution}}$$

Because of the division of soil P as discussed above, three methods of analysis will be presented in detail: (1) determination of total P, (2) determination of organic P, and (3) determination of available P.

10.2.2 Determination of Total P

The *fluoro-boric acid digestion procedure* (Bernas, 1968), employing specially designed acid digestion vessels, called bombs (Parr Instrument Co., Moline, IL), is presented here because of its rapidity, simplicity, and the need of only small amounts of reagents. The traditional method by fusion with NaOH or Na_2CO_3 needs expensive Pt crucibles and a variety of chemicals. The digest obtained by the bomb procedure can also be used for analysis of other elements (total elemental analysis of soil and plant samples).

Procedures

Reagents

1. Hydrofluoric acid, HF, 48%
2. Aqua regia: Mix 1 part of concentrated HNO_3 with 3 parts of concentrated HCl. The present author noticed that this reagent can be deleted without interfering in the digestion process.
3. Powdered boric acid, H_3BO_3
4. Ammonium molybdate solution: Weigh 25 g [$(NH_4)_6Mo_7O_{24} \cdot 4H_2O$] in a beaker, and add 200 mL distilled water. Heat at 60°C to dissolve under constant stirring. The solution is filtered to remove any coarse residue. Dilute 275 mL concentrated H_2SO_4 (cp grade) with distilled water to 750 mL, and allow the solution to cool. The molybdate solution is then poured slowly into the sulfuric acid solution.
5. Antimony: Dissolve 667.0 mg of potassium antimony tartrate, $KSbO \cdot C_4H_4O_6$, in 250 mL distilled water.
6. Ascorbic acid: Weigh 10 g of ascorbic acid in a 100-mL volumetric flask. Add 80 mL distilled water to dissolve, and dilute to the mark. Store this reagent in a cold room at 2°C.
7. Mixed reagent: Mix equal volumes of the ascorbic acid and the Sb solution before use. Prepare a fresh solution when required.
8. Standard solution of KH_2PO_4: Weigh 219.4 g of KH_2PO_4 in a 1-L volumetric flask, and dissolve with 500 mL distilled water. Add 25 mL 7 N H_2SO_4, and dilute to the mark with dis-

Determination of Macroelements

tilled water. This stock solution contains 50 mg P/L. Pipet 20 mL of this stock solution into a 500-mL volumetric flask, and dilute to the mark with distilled water. This is the working solution, which contains 2 µg P/mL.

Digestion and Extraction

Since HF is used in the digestion procedure, polyethylene beakers, stirrers, and volumetric flasks have to be employed in collecting and processing the results.

Weigh 50 mg sample, ground to 150 mesh, into the teflon vessel of the acid digestion bomb (20 mg for samples high in total P). After adding 0.5 mL aqua regia and 3.0 mL HF (48%), the vessel is closed and reassembled in the stainless steel body of the bomb. The bomb is sealed by turning the knurled cap, placed in an oven, and heated at 110°C for 40 min. Cool and unscrew the bomb. Wash the content with 4-6 mL distilled water by spraying into a 100- mL polyethylene beaker. Washing should not exceed 10 mL. Take care to wash all precipitates into the beaker. Add 2.8 g H_3BO_3, and stir with a plastic stirrer to dissolve the boric acid. Add 5-10 mL H_2O and any precipitate should dissolve at this time. Continue adding distilled water to about 40 mL. Transfer the solution into a 100-mL polyethylene volumetric flask, make up to volume, and store the solution in a plastic vial for further analysis. This solution can be used for total P determination and other macro-, and microelement analyses.

Ascorbic Acid Sulfomolybdo-Phosphate Blue Color Analysis of P

Pipet 5 mL of the digest into a 50-mL polyethylene volumetric flask. Aliquots of <2 mL or >2 mL can also be used, depending on the intensity of the blue color developed. Add 10 mL of the ammonium molybdate solution containing H_2SO_4 and swirl the flask to mix. Add 4 mL of the antimony-ascorbic acid mixture, and dilute with distilled water to the mark. Maximum color intensity develops within 10 min, and remains stable for several hours. The absorbance is measured at 840 nm.

Prepare a standard curve by pipetting 0, 1, 4, 5, and 10 mL of the KH_2PO_4 working solution into a series of 50-mL volumetric flasks. Proceed according to the procedure described above.

10.2.3 Determination of Organic P

Organic P can be determined indirectly by (1) sequential extraction, or (2) ignition methods (Jackson, 1960; Olsen and Sommers, 1982).

Procedures

1. *Sequential extraction method*

This method is originally suggested by Mehta et al. (1954), and involves extraction of inorganic P by HCl and NaOH, followed by extraction of total P with $HClO_4$. The organic P content is then found by difference:

$$P_{organic} = \text{total P } (HClO_4) - \text{inorganic P } (HCl+NaOH)$$

However, in the author's opinion, total P can be determined more accurately by the fluoro-boric acid bomb method as described above. Therefore, the $HClO_4$ digestion step in the sequential extraction method can be deleted, if the user wants to do so. This step is provided here only for completeness and for the purpose to provide the reader with an alternative method.

Reagents
1. Hydrochloric acid, HCl, concentrated
2. 0.5 N NaOH
3. 5 N NaOH

Determination of Macroelements

4. Perchloric acid, $HClO_4$, 72%
5. p-nitrophenol, 0.25% (wt/vol)
6. Reagents for P analysis in determination of total P

Extraction of Total and Inorganic P

Weigh 1 g of soil in a 100-mL polyethylene centrifuge tube and add 10 mL of concentrated HCl. Mix by swirling, and heat the tube at 70°C for 10 min in a waterbath. Remove the tube, and add an additional 10 mL concentrated HCl. Mix the content, and allow it to stand at room temperature for 1 h. Add 50 mL of distilled water, and centrifuge the suspension at 2000 rpm. Collect through filtration the clear supernatant in a 250-mL volumetric flask.

Add to the remaining sample in the centrifuge tube 30 mL of 0.5 N NaOH, mix, and allow the suspension to stand for 1 hour. Centrifuge the soil suspension, and filter the supernatant (containing dissolved organic matter) into the 250-mL volumetric flask containing the HCl extract. Add 60 mL 0.5 N NaOH to the remaining soil, mix thoroughly, and heat the tube at 90°C for 8 h in an oven. Centrifuge the suspension after cooling, and filter the supernatant into the 250-mL volumetric flask containing the HCl and cold-NaOH extracts. The combined extracts, representing total P, are diluted to volume with distilled water, and mixed thoroughly.

Pipet 10 mL of the thoroughly suspended material into a 100-mL beaker. Add 1 drop of concentrated H_2SO_4 followed by 1 mL of 72% $HClO_4$, and heat the mixture on a hot plate. Raise the temperature slowly until white fumes of $HClO_4$ appear. Cover the beaker with a watch glass, and continue digestion for another 30 min at 205°C. Cool the beaker and rinse quantitatively the digest with distilled water into a 50-mL volumetric flask, after which the volume is made up to the mark. This solution is used for determination of *total P*.

The organic matter fraction in the combined extract is then allowed to flocculate and settle out. The clear supernatant liquid is used for determination of *inorganic P*.

Analysis of Total and Inorganic P
Pipet 2 mL of either digest into a 50-mL volumetric flask. Add 5 drops of 0.25% p-nitrophenol and neutralize with NaOH until a yellow color persists. Dilute to 30 mL with distilled water, and proceed according to the sulfomolybdo-phosphate blue color method as described above. Prepare a blank containing all reagents, and a standard curve.

2. **Ignition method**
The organic P content is determined by extraction before and after ignition of the sample. Inorganic P is extracted from samples before ignition. Because of ignition organic P is converted into inorganic P, and total inorganic P is extracted from ignited samples. The difference between these two measurements is organic P:

$$P_{org} = \text{total P (ignited)} - \text{inorganic P (nonignited)}$$

Reagents
1. H_2SO_4, 1 N
2. NaOH, 5 N
3. p-Nitrophenol, 0.25% (wt/vol)
4. Reagents for analysis of P in section 10.2.2

Ignition of Sample
Weigh 1 g of soil in a porcelain crucible, and heat the sample slowly in a muffle furnace by gradually raising the temperature to 550°C over a period of 1 h. Heat the sample for 1 h at 550°C, and allow it to cool to room temperature.

Determination of Macroelements

Extraction of Ignited and Nonignited Samples

Transfer quantitatively the ignited soil sample into a 100-mL polyethylene centrifuge tube. Weigh in another tube 1 g of nonignited sample. Add to each of the tubes 50 mL 1 N H_2SO_4, and shake the mixture mechanically for 16 h. Centrifuge the samples for 15 min at 2000 rpm, and filter the supernatants into 100-mL plastic vials for storage and further analysis.

Pipet 2 mL of extract from each vial into 50-mL volumetric flasks, add to each of them 5 drops of 0.25% p-nitrophenol, neutralize to yellow with 5 N NaOH, and proceed as discussed in the previous section for analysis of total and inorganic P.

10.2.4 Determination of Available P

Procedures

1. **Phosphorus soluble in water**

Reagents
1. Distilled water
2. Reagents for sulfomolybdo-phosphate blue color method

Extraction and Analysis of P

Weigh 5 g of soil in a centrifuge tube, add 50 mL distilled water, and shake mechanically for 10 min. Centrifuge at 2000 rpm for 15 min, and the clear supernatant is filtered into a 100-mL volumetric flask. Dilute to the mark with distilled water.

Pipet 2 mL (5 mL for samples low in P) of the extract into a 25-mL volumetric flask, and proceed according to the procedure for P analysis by the sulfomolybdo-phosphate blue color method.

2. *Phosphorus soluble in dilute HCl+NH₄F or Bray 1 method*

Reagents
1. HCl, 0.5 N: Dilute 20.2 mL concentrated HCl in a 500-mL volumetric flask to volume with distilled water.
2. NH_4F, 1 N: Weigh 37 g NH_4F in a 1-L polyethylene volumetric flask, dissolve with 800 mL, and dilute to the mark with distilled water.
3. Extracting solution: Mix 15 mL of NH_4F solution with 25 mL of HCl, and dilute with distilled water to 460 mL. The mixture can be used with pyrex glassware.
4. Reagents for analysis of P by the sulfomolybdo-phosphate blue color method.

Extraction and Analysis of P

Weigh 1 g of soil in a polyethylene centrifuge tube, add 10 mL of extracting solution mixture, and shake mechanically for 10 min. Centrifuge, and filter the supernatant into a 25-mL volumetric flask. Dilute the solution to the mark with distilled water.

Pipet 2 mL of the extract into a 50-mL volumetric flask, and proceed according to the sulfomolybdo-phosphate blue color method.

3. *Phosphorus soluble in NaHCO₃ or Olsen method*

Reagents
1. $NaHCO_3$, 0.5 M: Weigh 42 g $NaHCO_3$ in a 1-L Pyrex beaker. Add 800 mL distilled water to dissolve, and adjust the pH to 8.5 with 1 M NaOH under constant stirring. Transfer the solution into a 1-L volumetric flask, and dilute to the mark. Store the solution in a polyethylene container. Prepare a fresh solution if it has been standing for a month.
2. H_2SO_4, 5 N: Measure 141 mL concentrated H_2SO_4 into a 1-L

volumetric flask. Add 500 mL water, cool the solution, and dilute to the mark with distilled water.
3. Reagents for analysis of P by the sulfomolybdo-phosphate blue color method.

Extraction and Analysis of P
Weigh 5 g of soil in a 250-mL Erlenmeyer flask, add 100 mL of the $NaHCO_3$ solution, and shake the flask mechanically for 30 min. Centrifuge the suspension at 2000 rpm for 15 min, and filter the supernatant into a 200-mL volumetric flask. Dilute with distilled water to the mark.

Pipet 5 mL of the clear extract into a 25-mL volumetric flask, adjust the pH with H_2SO_4 to pH 5.0. The amount of H_2SO_4 needed is determined by pipetting 5 mL of $NaHCO_3$ in a beaker and titrating it with 5 N H_2SO_4 to pH 5 under constant stirring. This amount of H_2SO_4 is then measured and added to the $NaHCO_3$ extract in the 25-mL volumetric flask, and proceed according to the sulfomolybdo-phosphate blue color method.

4. *Phosphorus soluble in dilute $HCl+H_2SO_4$, Mehlich's No.1 method, or double acid procedure*

Reagents
1. Double acid mixture: Measure 12 mL concentrated HCl and 73 mL concentrated H_2SO_4 (cp grade) and dilute to 15 L with distilled water. Cool, and add additional water to adjust the volume to 18 L. This mixture contains 0.05 N HCl and 0.025 N H_2SO_4.
2. Reagents for analysis of P by the sulfomolybdo-phosphate blue color method.

Extraction and Analysis of P
Weigh 5 g of soil in a 100-mL polyethylene centrifuge tube, add 200 mg activated charcoal (Darco G 60), and 20 mL of the double acid mixture. Shake the mixture mechanically for 10

min, centrifuge at 2000 rpm for 15 min, and filter the supernatant into a 50-mL volumetric flask. Dilute to volume with distilled water.

Pipet 5 mL of the extract into a 25-mL volumetric flask, and proceed according to the sulfomolybdo-phosphate blue color method.

The double acid extract can also be used for determination of other macro- and microelements content.

5. *Phosphorus soluble in NH_4HCO_3+DTPA*

Reagents

Extracting reagent: Weigh 1.97 g DTPA (diethylenetriaminepentaacetic acid) in a Pyrex beaker and add 800 mL of distilled water to dissolve under constant stirring. Add 2 mL of (1:1) NH_4OH to facilitate the dissolution process of DTPA. Add 79.06 g of NH_4HCO_3 under constant stirring until it dissolves. Adjust the pH of the mixture to 7.6 by titrating with NH_4OH. Transfer the solution to a 1-L volumetric flask and dilute to volume with distilled water. The solution is very unstable, and should be stored under a thin layer of mineral oil.

Extraction and Analysis of P

Weigh 5 g of soil in a 100-mL polyethylene centrifuge tube, add 20 mL of extracting solution, and shake the tube mechanically for 15 min. Centrifuge the suspension, and filter the supernatant into a 25-mL volumetric flask. Dilute the extract to volume with distilled water.

Pipet 2 mL into a 25-mL volumetric flask, and proceed according to the sulfomolybdo-phosphate blue color method.

6. *Anion resin-extractable P*

Reagents

1. Anion exchange resin, Dowex-2 (strong-base type), particle

size >30-mesh and <50-mesh. Cl-saturated.
2. NaCl solution, 10%

Extraction and Analysis of P

Weigh 1 g of soil, ground to pass a sieve of 100-mesh, in a 250-mL Erlenmeyer flask. Add 1 g of resin and 10 mL of distilled water, and shake the mixture mechanically for 16 h. Separate the resin from the soil by sieving through a fine-meshed screen that retains only the resin. Transfer the resin from the sieve into a 50-mL Pyrex beaker containing 25 mL of NaCl solution (10%). Place the beaker in a waterbath for 45 min, cool the mixture, and filter the solution into a 100-mL volumetric flask. Wash or leach the resin with addition NaCl solution (10%) until 100 mL of filtrate is obtained.

Pipet 10 mL of the solution into a 25-mL volumetric flask, and proceed according to the sulfomolybdate-phosphate blue color method.

7. *Phosphorus determined by isotopic dilution of ^{32}P*

Reagents
1. ^{32}P: Dilute an aliquot of the ^{32}P with 5 µM KH_2PO_4 (5 µg/L) to obtain a suitable counting rate.
2. KH_2PO_4, 100 µM (100 µg/L), pH 7.0 (adjusted with KOH).
3. Toluene or chloroform
4. Reagents for analysis of P by the sulfomolybdo-phosphate blue color method.

Extraction and Determination of ^{31}P and ^{32}P

Weigh 5 g of soil in a 250-mL Erlenmeyer flask. Add 99 mL of distilled water, and shake at constant temperature for 1 h. Add 1 mL of the diluted ^{32}P solution, and 2 drops of toluene or chloroform. Shake the mixture mechanically for 24 h. Centrifuge at 10 000 rpm for 15 min to obtain a clear supernatant solution.

Pipet 2 mL of the supernatant, and determine the $^{31}P_{solution}$ (= $P_{solution}$) concentration by the sulfomolybdo-phosphate blue color method.

Pipet another aliquot of the supernatant for determination of ^{32}P by the liquid scintillation method.

Calculations of $^{31}P_{solid}$ and Labile ^{31}P

The concentration of $^{31}P_{solid}$ is calculated using the equation:

$$^{31}P_{solid} = (^{32}P_{solid}/^{32}P_{solution}) \times {}^{31}P_{solution}$$

The concentration of labile ^{31}P is calculated using the equation:

$$\text{Labile } ^{31}P = P_{solution}/f$$

in which f = fraction of ^{32}P left in solution.

10.3 DETERMINATION OF POTASSIUM

10.3.1 Forms of Potassium in Soils

Potassium, K, is an essential element for plant growth. In the fertilizers industry this element is called *potash*, a term derived from potassium salts obtained by burning wood in pots (*pot-ash*). Together with N and P, it is one of the major fertilizer elements.

Potassium is present in igneous, sedimentary, and metamorphic rocks. Its total content in basaltic and granitic rocks is on the average 2.2% expressed as K_2O, whereas its content in mineral soils is approximately 0.83% (Tan, 1994). The mineral sources for potassium are K-aluminum silicates, including potash-feldspars, e.g., orthoclase, and micas, e.g., muscovite and biotite. Potassium also occurs in the clay mineral illite, and in miscellaneous chloride, sulfate, and borate minerals. The dissolution of K from aluminum silicates is attributed to hydrolysis reactions, which are weathering processes. The chemical reaction can be represented as follows:

$$KAlSi_3O_8 + H_2O \rightarrow HAlSi_3O_8 + KOH$$
orthoclase clay soluble

The KOH dissolves in water and releases potassium in the form of K^+, a valuable plant nutrient. Generally the concentration of K^+ ions is relatively low compared with their total proportions in soils. On the average, the K^+ ion concentration in the soil solution is 5 mg/L, but this can increase considerably as a result of application of K fertilizers. The K^+ ion is a very stable ion, and is very difficult to precipitate. Therefore, when K^+ ions are not used by plants, they will be leached rapidly from the soil. Only by complex formation or cation exchange can K^+ ions be precipitated and immobilized.

As is the case with phosphorus, the total K content does not reflect its availability to plants, and efforts have been spent for developing the so-called *K-availability indices*. Methods were developed in the past to extract that portion of total K from soils that is available to plants (Knudsen et al., 1982). The NH_4OAc, double acid, Bray P1, Morgan's method, and other procedures have been used in the past to extract available K. According to the respective extraction procedures, several types of available K have been proposed (McLean and Watson, 1985):

1. *Exchangeable K*
This is the amount of K extracted from the exchange sites by the NH_4OAc, $MgCl_2$, $CaCl_2$, or NaCl method.

2. *Nonexchangeable K*

This fraction of soil K cannot be extracted by the above mentioned methods. This type of soil K is usually extracted by boiling soils with HNO_3. In the author's opinion, this fraction of soil K should be classified as *nonavailable* to plants. Plants do not have the capacity equivalent to that of boiling (hot) HNO_3 in extracting this fraction of soil K.

3. *Plant-available K*

This is the amount of K soluble in water, hence extractable with water. The name plant-available K is rather confusing for water soluble K, since plant-available K also includes exchangeable K. The present author suggests to change the name into *water-soluble K*. The latter is consistent with *phosphorus soluble in water* used in the nomenclature for available P.

4. *Fixed K*

This is the amount of K trapped in the intermicellar spaces of expanding minerals. The present author suggests to also delete this category, since fixed K is relatively nonavailable to plant growth. The release from fixed positions, which is referred to as *released K (5th category of available K)* by McLean and Watson (1985), depends on several factors, including especially saturation and desaturation of the soil with K.

Of the types of available K listed and discussed above, perhaps the most viable are (1) exchangeable K, and (2) water soluble K. To these should be added *double acid extractable K*. This dilute acid extractable K fraction is commonly accepted today as K available to plants (Jones, 1985).

Because of the above, the methods presented in this book includes only (1) determination of total K, and (2) determination of available K by the (a) NH_4OAc method, and (b) double acid procedure.

10.3.2 Determination of Total K

Procedures
Fusion of samples with solid NaOH or $NaCO_3$, and digestion using mixtures of concentrated H_2SO_4 and $HClO_4$ have been employed for total elemental analysis. However, as indicated in an earlier section, the fluoro-boric acid digestion method employing specially designed bombs (Bernas, 1968) is the most rapid and convenient method for total elemental analysis of soil, mineral, and plant samples. The procedures discussed below are also applicable to determination of Na.

Reagents
1. Hydrofluoric acid, HF, 48%
2. Aqua regia: mix 1 part of conc HNO_3 with 3 parts conc HCl
3. Powdered boric acid, H_3BO_3
4. Stock K solution containing 1000 µg K/mL (Fisher Scientific Co.). Pipet 2 mL from the stock solution into a 1-L volumetric flask, and dilute to the mark with distilled water. This solution, called the standard solution, contains 2 µg K/mL.

Extraction of Total K
The following procedure of the fluoro-boric acid digestion method has been discussed in the determination of total P (section 10.2.2). However, for reader's convenience, it is discussed again below with some minor revisions.

Polyethylene beakers, stirrers, and volumetric flasks have to be employed with this method, because HF is used.

Weigh 50 mg sample, ground to 150 mesh, into the Teflon vessel of the acid digestion bomb (20 mg for samples high in total K). After adding 0.5 mL aqua regia and 3.0 mL HF (48%), the vessel is closed and reassembled in the stainless steel body of the bomb. The bomb is sealed by turning the knurled cap, placed in an oven, and heated at 110°C for 40 min. Cool and unscrew the bomb. Wash the content with 4-6 mL distilled water by spraying into a 100-mL polyethylene beaker. Washing should not exceed 10 mL. Take care to wash all precipitates into the beaker. Add 2.8 g H_3BO_3, and stir with a plastic stirrer to dissolve the boric acid. Add 5-10 mL H_2O and any precipitate should dissolve at this time. Continue adding distilled water to about 40 mL. Transfer the solution into a 100-mL polyethylene volumetric flask, make up to volume, and store the solution in a plastic vial for analysis of major soil K and other macro-, and micro-elements.

Analysis of Total K by Atomic Absorption Spectroscopy

Calibrate the atomic absorption spectrophotometer, provided with a hollow cathode K tube, with the standard containing 2 µg K/mL. This standard should give a reading of about 0.22 absorbance units with the Perkin-Elmer instrument model 503. Other models require different standards and readings. Check your instrument's manual for the correct settings.

Aspirate your extract into the instrument and read the absorbance. If your reading is >0.22 absorbance, dilute your extract so that the reading is below or equals 0.22 absorbance. The working range for K is linear from 0 to 2 µg K/mL.

10.3.3 Determination of Available K

Procedures

1. ***NH_4OAc method***

Reagents
NH_4OAc, or ammonium acetate (CH_3COONH_4), 1 M, pH 7.0. See section 9.3.1 for the preparation of this solution.

Extraction of Exchangeable K

Weigh 5.0 g soil in a 100-mL polyethylene centrifuge tube and

Determination of Macroelements

proceed as described in section 9.3.1.

Analysis of Exchangeable K by Atomic Absorption Spectrophotometry
See the procedure for determination of total K described above.

2. ***Double acid procedure***
For the procedures and analysis of dilute acid extractable K, see the method for analysis of phosphorus soluble in dilute $HCl+H_2SO_4$, Mehlich's No.1, or double acid method in section 10.2.4.

10.4 DETERMINATION OF SODIUM

10.4.1 Forms of Sodium in Soils
Sodium is not considered an essential nutrient for plant growth. In the fertilizer industry it is called soda, and is present in nitrate of soda, $NaNO_3$.

Sodium compounds are widely distributed in nature. The major mineral source is Na aluminum silicates, also called *sodic feldspars* or *plagioclase*, e.g., *albite*, $NaAlSi_3O_8$. This mineral is a major components of basic igneous rocks. Sodium can also occur in nature as chloride, sulfate, and borate compounds. The dissolution of Na from albite is attributed to hydrolysis reactions, which is a weathering process. The chemical reaction is usually represented as follows:

$$NaAlSi_3O_8 + H_2O \rightarrow HAlSi_3O_8 + NaOH$$
albite   clay

Since NaOH is soluble in water, the sodium is released in the form of a Na$^+$ ion. Because of its cationic nature, the Na$^+$ ions are attracted by the negatively charged surfaces of soil colloids. They are then called *exchangeable Na*.

The total sodium content of normal soils is approximately 0.63%, but the Na$^+$ ion concentration in the soil solution of normal soils is lower than the total Na content stated above. On the average, the Na$^+$ ion concentration amounts to 10 mg/L or larger. High soluble Na contents are very harmful to the physical condition of soils. Sodium tends to disperse the soil colloids. A dispersed soil is puddled when wet and hard when dry.

Sodium ions are very stable in soil water, and are very difficult to precipitate. They may, however, react rapidly with Cl$^+$ to form NaCl, which can accumulate in arid regions or where drainage is inhibited. Accumulation of sodium salts in aridisols gives rise to crust formation on the surface soil and/or development of B_n horizons. In other soils, such an accumulation of NaCl contributes toward development of salinity, resulting in a decrease in the quality of soil and water. Because of this, the exchangeable Na content is used by the U.S. Salinity laboratory as a criteria for the distinction between saline and sodic soils. This Na content is expressed in terms of ESP (exchangeable sodium percentage), which is defined as follows:

$$ESP = (exch.Na^+/\Sigma\ exch.cations) \times 100\%$$

Saline soils are characterized by an ESP <15%, and an electrical conductivity, EC, >4 mmho/cm (25°C). For more details on the use of ESP and EC in soil classification, reference is made to Tan (1994), and Soil Survey Staff (1975).

Because Na is not an essential plant nutrient, no effort has been made to assess its availability to plants. As indicated above, the Na content is found useful only in soil characterization, determination of CEC, and percentage base saturation, and soil genesis and classification of especially aridisols. Therefore, only the following methods of determination of Na will be discussed below: (1) determination of total Na, and (2) determination of exchangeable Na.

10.4.2 Determination of Total Na

Procedures

Fluoro-Boric Acid Digestion Method
Fusion of samples with solid NaOH or $NaCO_3$, as suggested for determination of total K, cannot be used in the determination of Na. Digestion using mixtures of concentrated H_2SO_4 and $HClO_4$ as for total elemental analysis is a possibility. However, the most convenient method for determination of total Na content in soils is the fluoro-boric acid digestion method employing specially designed bombs (Bernas, 1968). For the procedure, reagents, and extraction of total Na by the fluoro-boric acid method reference is made to section 10.3.2.

Analysis of Total Na by Atomic Absorption Spectroscopy
Calibrate the atomic absorption spectrophotometer, provided with a hollow cathode Na tube, with a standard containing 0.8 µg Na/mL. The standard is prepared by pipetting 0.8 mL from of stock solution, containing 1000 µg Na/mL (Fisher Scientific Co., Atlanta, GA), into a 1-L volumetric flask, and diluting it to volume with distilled water. This standard should give a reading of about 0.23 absorbance units with the Perkin-Elmer instrument model 503. Other models require different standards and readings. Check your instrument's manual for the correct settings.

Aspirate your extract into the instrument and read the

absorbance. If your reading is >0.23 absorbance, dilute your extract so that the reading is below or equals 0.23 absorbance. The working range for Na is linear from 0 to 1 µg Na/mL.

10.4.3 Determination of Exchangeable Na

Procedure

NH$_4$OAc method

Reagents
NH$_4$OAc, or ammonium acetate (CH$_3$COONH$_4$), 1 M, pH 7.0. See section 9.3.1 for the preparation of this solution.

Extraction of Exchangeable Na
Weigh 5.0 g soil in a 100-mL polyethylene centrifuge tube and proceed as described in section 9.3.1.

Analysis of Exchangeable Na by Atomic Absorption Spectrophotometry
See the procedure for analysis of total Na by atomic absorption as described above.

10.5 DETERMINATION OF CALCIUM

10.5.1 Forms of Calcium in Soils

Calcium, Ca, is a very important cation in soils, soil water, and waters in lakes and streams. In addition to increasing pH of acid soils, it is believed to have a beneficial effect in development of soil structure. It is also an essential macronutrient for plant growth.

The primary mineral sources of Ca are calcite, aragonite, dolomite, and gypsum. Calcite, CaCO$_3$, is the major constituent of limestone, calcareous marls, and calcareous sandstone (Tan,

Determination of Macroelements

1994). The total Ca content in soils is estimated to be 1.4%. Depending on climatic conditions and parent material, the Ca content may vary considerably from soil to soil. Soils in desert climates may be high in Ca, often containing Ca as $CaCO_3$ nodules in the Bk horizon. In humid region soils, drastic leaching has removed most of the Ca minerals from the soils. Therefore, soils in humid regions, e.g. alfisols, ultisols, and oxisols, are relatively low in Ca. They are often acid in reaction.

Calcium carbonate is insoluble in water, but will dissolve in water containing CO_2, a process called *carbonation*, which is a weathering process. The reaction can be written as follows:

$$H_2O + CO_2 \rightarrow H_2CO_3$$
$$\text{carbonic acid}$$

$$H_2CO_3 + CaCO_3 \rightarrow Ca(HCO_3)_2$$
$$\text{calcium bicarbonate}$$

The calcium bicarbonate is soluble in water and releases the Ca^{2+} ion, which is an important plant nutrient. In the presence of negatively charged clay, the Ca^{2+} ions are adsorbed by the clay and are called exchangeable Ca. The Ca concentration in soil solutions is reported to vary from 1.7 mM to 19.4 mM (Clark, 1984; Adams, 1974). A Ca concentration between 0.1 to 1 mM at the root surface is considered adequate for plant growth. However, a Ca concentration between 1 to 5 mM is required to protect plant roots against the harmful effect of low pH and its consequences.

In view of the discussion above, the methods for determination of Ca include (a) determination of total Ca, (b) determination of exchangeable Ca, and (c) determination of dilute acid extractable Ca. Although the latter is considered nonexchangeable Ca, it is reported to be also available to plant growth.

10.5.2 Determination of Total Ca

The procedures and determination of total Ca are exactly the same as those for total potassium. The extract for determination of K obtained by the fluoro-boric acid method is also used for total Ca analysis by atomic absorption spectrophotometry. In this case, the instrument, provided with a hollow cathode Ca tube, is calibrated with a standard solution containing 4 µg Ca/mL, which should give an absorbance reading of 0.20 units with a Perkin-Elmer AA spectrophotometer, model 503. The working range of Ca is linear from 0 to 4 µg Ca/mL. The standard solution is prepared by pipetting 4 mL of a stock solution (1000 µg Ca/mL), purchased from Fisher Scientific Co., into a 1-L volumetric flask, and diluting it to the mark with distilled water.

10.5.3 Determination of Exchangeable Ca

The same NH_4OAc extract for the determination of exchangeeable cations (see section 9.3.1) is used for the determination of exchangeable Ca by atomic absorption spectrophotometry. The parameters for calibration of the instrument remains the same and are discussed above in section 10.4.2.

10.5.4 Determination of Dilute Acid Soluble Ca

Calcium is extracted by the double acid procedure as described above in section 10.2.4. That same extract is analyzed for Ca concentration by atomic absorption spectrophotometry using similar parameters and a calibration procedure as indicated in section 10.4.2.

10.6 DETERMINATION OF MAGNESIUM

10.6.1 Forms of Magnesium in Soils

Magnesium, Mg, is one of the essential macronutrients for plant growth. It is an essential constituent of chlorophyll, vital in photosynthesis.

The minerals containing Mg are dolomite, Mg silicates, Mg phosphates, Mg sulfides, and Mg molybdates. Dolomite, $CaMg(CO_3)_2$, a major constituent of dolomitic limestone, is the most common source of Mg in soils. It is a mineral found in sedimentary rocks (Tan, 1994).

The dissolution of dolomitic limestone is affected by the CO_2 content in soil water and the reaction is represented as follows:

$$CaMg(CO_3)_2 + H_2CO_3 \rightarrow Ca^{2+} + Mg^{2+} + 4HCO_3^-$$

The Ca^{2+} and Mg^{2+} released by the above reaction are available for plant growth. Since they are cations, they will be adsorbed by negatively charged clay surfaces, and are then called exchangeable Ca and Mg. Exchangeable Mg is also available to plants.

The average Mg content in soils is approximately 0.5%, whereas its concentration in soil water is estimated to be 10 mg/L. In soil water, Mg exists as the cation Mg^{2+}, or as ion pairs with HCO_3^-, SO_4^{2-}, and Cl^-.

Because Mg is present in several forms in soils as discussed above, three procedures for determination of Mg will be presented below: (1) determination of total Mg, (2) determination of exchangeable Mg, and (3) determination of dilute acid extractable Mg. As is the case with Ca, dilute acid soluble Mg is also available to plants.

10.6.2 Determination of Total Mg

The extract obtained by the fluoro-boric acid digestion procedure for total K determination (section 10.3.2) is used for determination of total Mg by atomic absorption spectrophotometry. The instrument, provided with a hollow cathode Mg tube, is calibrated with a standard solution containing 0.3 µg Mg/mL, which gives an absorbance reading of 0.19 units with a Perkin-Elmer AA spectrophotometer, model 503. The working range of Mg is linear from 0 to 0.5 µg/mL. The standard solution is prepared in two steps as follows. In the first step: pipet 3 mL of stock solution (1000 µg Mg/mL), purchased from Fisher Scientific Co., into a 1-L volumetric flask. Dilute to volume and a solution containing 3 µg Mg/L is obtained. In the second step: pipet 10 mL of this solution, containing 3 µg Mg/mL, into a 100-mL volumetric flask and dilute to volume with distilled water. This solution, containing 0.3 µg Mg/mL, is the standard solution for calibration of the AA instrument.

10.6.3 Determination of Exchangeable Mg

The NH$_4$OAc extract obtained for determination of exchangeable cations (section 9.3.1) is also used for the determination of exchangeable Mg. The parameters and calibration of the AA spectrophotometer are discussed in the preceding section (10.5.2).

10.6.4 Determination of Dilute Acid Soluble Mg

The extract obtained by the double acid procedure (see section 10.2.4, procedure No.4) is also used for the determination of dilute acid extractable Mg by atomic absorption spectrophotometry. The parameters and calibration of the AA spectrophotometers are discussed in section 10.5.2.

10.7 DETERMINATION OF SULPHUR

10.7.1 Forms of Sulphur in Soils

Sulphur, S, is a major macronutrient for plants. It is an essential constituent of vitamin B, thiamine, biotin, and several amino acids, such as cysteine, cystine, and methionine. Almost 90% of the S in plants are present in the form of these amino acids. The entry of S into the plant body is conditioned by the reaction of SO_4^{2-} with ATP (adenosine triphosphate) to form adenosine phosphosulfate (Stevenson, 1986).

Sulfur is present in soils in organic and inorganic form. Organically, S is an important constituent of proteins and amino acids, as indicated above. The major inorganic sources of S include gypsum, $CaSO_4$, and pyrite, Fe_2S. Sulphur can also be added to soils by the use of chemicals containing S, such as insecticides and detergents. It exists in soils and soil water mainly as the SO_4^{2-} anion in combination with the cations Ca^{2+}, Mg^{2+}, K^+, Na^+, or NH_4^+. Present in the form of elemental sulfur, S, it will be oxidized in aerobic condition and converted quickly into SO_4^{2-}. Under anaerobic condition, SO_4^{2-} will be reduced by microorganisms into SO_3^{2-} or H_2S. Hydrogen sulfide is formed especially in swamps and other areas with stagnant water. Waterlogging in coastal regions and paddy soils provide a suitable environment for formation of H_2S. In soils rich in Fe, H_2S is usually precipitated as Fe_2S, which imparts to soils a black color, as is frequently noted in soils of the coastal regions and tidal marshes. When H_2S is allowed to accumulate, it is not only toxic to soil organisms, but it also creates environmental problems. A group of bacteria present in soils is capable of oxidizing the H_2S into elemental S and SO_4^{2-}. The reactions can be written as

$$2H_2S + O_2 \rightarrow 2S + 2H_2O$$

$$2S + 2H_2O + 3O_2 \rightarrow 2H_2SO_4$$

The formation of sulfuric acid, H_2SO_4, will decrease soil pH to 2.0, resulting in formation of *acid sulphate soils*, sometimes also called *cat-clays*. In addition to the toxicity created by the extremely low pH, the acidity will liberate toxic amounts of Al and Fe.

The total S content in soils varies widely from soil to soil. Sandy soils in the humid regions are generally low in S (0.002%). In contrast, soils in arid regions may contain 5% S. In general, the total S content in agricultural soils of humid and semihumid regions ranges from 0.01 to 0.05%, which is equivalent to 224-1120 kg S/ha (or 200-1000 lbs S/acre). The data in Table 10.3 illustrates the approximate S content of the various forms of S in a mollisol.

Plants absorb S mostly in the sulphate, SO_4^{2-}, form. Consequently the sulfate form is called *available S*. The anionic nature of SO_4^{2-} prevents its attraction by clay colloids, although some soil sesquioxide minerals have been reported to adsorb considerable amounts of SO_4^{2-}. Consequently SO_4^{2-} is expected to be lost rapidly from soils by leaching.

Therefore two types of S determination will be discussed in this book: (1) determination of total S, and (2) determination of inorganic S. The organic S fraction can be determined by difference:

Organic S = total S - inorganic S

Table 10.3 The Average Organic and Inorganic S Content in a Mollisol (Stevenson, 1986).

Form of S	as µg/g of soil	in % of total S
Total	292	100
Organic	283	97
Inorganic	8	3
Sulfate	8	3

Inorganic S, as indicated above, is composed of mainly SO_4. It occurs in the soil as free, and adsorbed SO_4^{2-} ions, and as salts, mainly in the form of gypsum ($CaSO_4$). Of these three, *free* and *adsorbed* SO_4^{2-} make up the majority, whereas gypsum is frequently present only in very small amounts in soils. Other SO_4-compounds, such as jarosite [$KFe_3(OH)_6(SO_4)_2$], are even more uncommon in soils than gypsum. If they are present, they constitute an insignificant fraction of the total SO_4 fraction in soils.

The determination of total S can be performed by a variety of methods. But all of them are based on the oxidation of S into sulfate, SO_4, e.g., digestion with $HClO_4$, HNO_3, H_3PO_4, HCl or a mixture of these acids, by fusion with Na_2CO_3, or by dry ashing procedures (Tabatabai, 1982). None of these methods appear to be satisfactorily. The present author believes that the fluoro-boric acid digestion (see section 10.3.2) is perhaps a better method than the fusion method with Na_2CO_3. However, as an alternative choice the procedure of oxidation of S with $Mg(NO_3)_2$ (Hesse, 1972) is presented below.

10.7.2 Determination of Total S

Procedure

Reagents
1. $Mg(NO_3)_2$, 2 M: Dissolve 148 g magnesium nitrate in a 500-mL volumetric flask with 400 mL distilled water under constant stirring. Dilute the mixture to volume with distilled water.
2. HNO_3, 25%
3. Acetic acid, 50%
4. Ortho-H_3PO_4, concentrated
5. $BaCl_2$ crystals: the crystals are ground to pass a 0.5 mm sieve, and retained on a 0.25 mm sieve.
6. Gum acacia solution: 0.25% (w/v) in water.
7. Standard sulphate solution: Weigh 53.8 mg $CaSO_4.2H_2O$ into a 100-mL volumetric flask, add 80 mL water to dissolve under constant stirring. Make up to volume with distilled water. This solution contains 300 µg SO_4/mL or 100 µg S/mL. The conversion factor of SO_4 into S is 0.3333.

Digestion and Conversion into SO_4^{2-}
Weigh 1.0 g of soil (ground to pass a 0.5 mm sieve) in a porcelain crucible, add 2 mL of $Mg(NO_3)_2$ solution, and heat the mixture at 70°C on a hotplate to dryness. Place the crucible containing the dry mixture in a muffle furnace and heat at 300°C for 12 h (overnight). Cool, add 5 mL of 25% HNO_3, cover the crucible and digest the mixture on a waterbath for 3 h. Avoid loss of acid. After cooling, dilute the mixture with 10 mL of distilled water and filter it into a 50-mL volumetric flask. Make to volume with distilled water.

Colorimetric Determination of SO_4^{2-}
Pipet 5 mL of the extract into a 50-mL volumetric flask, add 5 mL of acetic acid, 1 mL of H_3PO_4, and 1 g of $BaCl_2$ crystals.

Determination of Macroelements

The phosphoric acid will decolorize any Fe present in solution. Mix gently by inverting the flask several times. Add 2 mL of gum acacia solution, and make up to volume with distilled water. Mix gently again before measuring SO_4^{2-} concentration with a spectrophotometer at 490 nm. Prepare a calibration curve by pipetting 0, 1, 3, and 5 mL of the standard sulphate solution into a series of 50-mL volumetric flasks, and proceeding in the same manner as described above.

10.7.3 Determination of Inorganic S

Since most of the inorganic S is in the form of free (soluble) and adsorbed SO_4, procedures pertaining the determination only of free and adsorbed SO_4 are presented below. Free SO_4 is extractable with water, NaCl, $CaCl_2$, and LiCl solutions. Adsorbed SO_4 is usually extracted by exchange procedures using NH_4OAc, $NaHCO_3$, KH_2PO_4, or $Ca(H_2PO_4)_2$. The Morgan solution and the Bray P-1 solution are often suggested as extractants of adsorbed SO_4. The extraction procedures using $Ca(H_2PO_4)_2$ and $CaCl_2$ are presented below (Tabatabai, 1982) for the determination of free + adsorbed SO_4, and free SO_4, respectively. Measurements of the extracted SO_4 are, however, based on the colorimetric procedure (Hesse, 1972).

Procedure

Reagents
1. $Ca(H_2PO_4)_2 \cdot H_2O$ solution: Weigh 2.03 g of $Ca(H_2PO_4)_2 \cdot H_2O$ into a 1-L volumetric flask, add 800 mL distilled water to dissolve under constant stirring, and make up to volume with distilled water. This solution contains 500 mg P/L.
2. $CaCl_2$ solution, 0.15%: Weigh 1.5 g of $CaCl_2 \cdot 2H_2O$ into a 1-L volumetric flask, and add 800 mL distilled water. Dissolve it under constant stirring, and make up to volume with distilled water.

3. Reagents for colorimetric SO_4^{2-} determination described in section 10.7.2.

Extraction of Inorganic SO_4^{2-}

1. *Extraction of free and adsorbed SO_4^{2-}*
Weigh 5 g of soil into a 100-mL polyethylene centrifuge tube, add 50 mL of $Ca(H_2PO_4)_2$, and shake the mixture for 30 min with a wristaction shaker.

2. *Extraction of free (soluble) SO_4^{2-}*
Shake 5 g of soil with 50 mL of $CaCl_2$ solution. Centrifuge after shaking, and filter the supernatant into a 100-mL volumetric flask. Dilute to volume with distilled water

Colorimetric Determination of SO_4^{2-}
Pipet 5 mL of the extract into a 50-mL volumetric flask, add 5 mL of acetic acid, 1 mL of H_3PO_4, and 1 g of $BaCl_2$ crystals. Proceed as described above in 10.7.2.

Chapter 11

DETERMINATION OF MICROELEMENTS

11.1 DETERMINATION OF ALUMINUM IN SOILS

11.1.1 Forms of Aluminum in Soils

Aluminum, Al, is an important constituent of soils. It is the most abundant element in rocks and minerals, and is released in soils upon weathering of these rocks and minerals (Tan, 1994). It is chemically very active, hence the element will seldom occur free in nature, and is usually present as oxides or as a constituent of alumino-silicate minerals and sesquioxide clays. The most important form of aluminum oxide is bauxite, $Al_2O_3 \cdot 2H_2O$, and gibbsite, $Al(OH)_3$, a sesquioxide clay. More exotic forms of aluminum oxides are corundum, Al_2O_3, ruby, and sapphire. Ruby is a red variety of corundum, whereas sapphire is a blue colored species of corundum. All three varieties belong to the hematite group of minerals, characterized by the formula X_2O_3, in which X represents Al or Fe. If X=Fe, the mineral is hematite, Fe_2O_3. In soils, Al can also occur as amorphous compounds, as coatings on soil minerals, and as organic complexes.

The total aluminum content in igneous rocks, in terms of Al_2O_3, is on the average 16.3 % (Tan, 1994). The Al_2O_3 content in laterites, highly weathered tropical soils, may range from 9% to 16% (Mohr and Van Baren, 1960). In contrast, the concentration of soluble aluminum, Al^{3+}, in soil water is very small and amounts to only 1.0 mg Al/L. In soils, the Al^{3+} ion is quickly attracted to negatively charged clay surfaces, and is called *exchangeable Al*. When an aluminum compound dissolves in soil water, the Al^{3+} ion is quickly surrounded by 6 molecules of H_2O in octahedral coordination: $Al(H_2O)_6^{3+}$. This ion, called an *aluminum hexahydronium ion*, is subject to hydrolysis, especially as soil pH increases:

$$Al(H_2O)_6^{3+} \leftrightharpoons Al(H_2O)_3(OH)_3^{\circ} + 3H^+$$

For simplicity the reaction above is usually written without the coordinated water:

$$Al^{3+} + 3H_2O \leftrightharpoons Al(OH)_3^{\circ} + 3H^+$$

Al is officially not considered a plant nutrient, although plants develop necrotic spots when grown in nutrient solutions deficient in Al. However, present in large amounts, these Al^{3+} ions may cause toxicity to plants. Al^{3+} ion concentrations between 50-100 mg/L have been reported to reduce plant growth considerably (Foy, 1978; Tan and Binger, 1986). Aluminum also plays an important role in creating soil acidity, and in addition reacts quickly with phosphates to form insoluble phosphate compounds, e.g. variscite (Tan, 1994). Such a reaction between Al and phosphate is called *phosphate fixation*, and may cause soluble P concentration to decrease in soils creating P deficiency to plants.

The Al concentration in soils is frequently used as an

Determination of Microelements

important indicator for degree of weathering, and as an important parameter in characterization of soils. The effective CEC (ECEC), a basic parameter in soil's characterization, is defined as the sum of KCl-extractable Al and NH_4OAc-exchangeable bases (Soil Survey Staff, 1975).

Because of the importance of Al in soil characterization, soil genesis, and soil classification, three types of determination of Al will be presented below: (1) determination of total Al, and (2) determination of exchangeable Al by the NH_4OAc method, and (3) determination of Al by the KCl extraction method. Exchangeable Al is sometimes considered synonymous to KCl-extractable Al (Barnhisel and Bertsch, 1982). Determination of available Al, as discussed in many other books, will not be presented here. As pointed out above, officially Al is not considered a plant nutrient. However, if *available Al* is a viable term, then the amount of Al extractable by NH_4OAc can be used as an index of availability to plants.

11.1.2 Determination of Total Al

Procedures

Fluoro-Boric Acid Digestion Method

The fluoro-boric acid digestion procedure (see sections 10.2.2 and 10.3.2) is suggested for the extraction of total Al from soils, because of its simplicity, rapidity, and accuracy in analysis. The use of this procedure for total elemental analysis will also make this book easy to read, and more consistent in analytical methods.

Reagents
1. Hydrofluoric acid, HF, 48%
2. Powdered boric acid, H_3BO_3
3. Stock solution containing 1000 µg Al/Ml (Fisher Sci. Co.).

Pipet 5 mL from the stock solution into a 100-mL volumetric flask, and dilute to the mark with distilled water. This solution, called the *standard solution*, contains 50 µg Al/mL, and is used for the determination of Al by atomic absorption spectroscopy.
4. Standard solution for colorimetric determination of Al:
   a. Aluminon reagent, 0.04%: Weigh 200.0 mg of aluminon (aurin tricarboxylic acid; Eastman Kodak Co., Rochester, NY) in a 100-mL volumetric flask, and dissolved in a buffer solution of pH 4.2.
   b. Buffer solution, pH 4.2: Dilute 60 mL of HOAc to 900 mL with distilled water in a 1-L volumetric flask. Add 90 mL NaOH (10%), adjust the pH to 4.2, and make up to volume with distilled water.
   c. Standard solution for colorimetric determination of Al: Pipet 5 mL of the Al stock solution (1000 mg/mL) into a 1-L volumetric flask, and dilute to volume with distilled water. This solution, called the standard solution, contains 5 µg Al/mL.

Extraction of Total Al
For the extraction procedure see sections 10.2.2. and 10.3.2.

1. *Analysis of total Al by atomic absorption spectroscopy*
The analysis of Al in the digest can be performed colorimetrically or by atomic absorption spectrophotometry. The latter is currently believed to be a more rapid and accurate method than the conventional colorimetric method. Therefore, the method using atomic absorption spectrophotometry will be discussed below.

Calibrate the atomic absorption spectrophotometer, provided with a hollow cathode Al tube, with the standard solution containing 50 µg Al/mL. This standard should give a reading of about 0.22 absorbance units with the Perkin-Elmer instrument model 503. Other models may require different standards and readings. Check your instrument's manual for the correct

Determination of Microelements

parameters.

Aspirate the fluoro-boric acid extract into the instrument and read the absorbance. If your reading is >0.22 absorbance, dilute your digest so that the reading is below or equals 0.22 absorbance. The working range for Al is linear from 0 to 50 µg Al/mL.

2. *Colorimetric determination of Al*

Pipet 5 mL of the digest (2 mL if high in Al content) into a 50-mL volumetric flask, add 10 mL of buffer solution, pH 4.2, and add distilled water to approximately 30 mL. Mixed by swirling, add 10 mL of aluminon solution, and adjust the volume to the mark. Mixed well, and measure after 25 min the intensity of the color at 520 mµ. Prepare a standard curve by pipetting 0, 2, 5, and 10 mL of the standard solution containing 5 µg Al/mL.

11.1.3 Determination of Exchangeable Al

Procedures

Two methods can be used for the determination of exchangeable Al, e.g., the NH_4-acetate method, and KCl method. Although both methods are suitable, some authors prefer to use the NH_4-acetate method, whereas others prefer the KCl method. In addition, KCl-extractable Al is used in the determination of ECEC. The two methods are presented below.

1. *NH_4OAc Method*

See sections 9.3.1 and 10.3.3 for reagents and extraction procedures.

For the analysis of exchangeable Al by atomic absorption spectroscopy see the preceding procedure in section 11.1.2.

2. *KCl Method*
1. KCl, 1 *M*: Weigh 59.0 g KCl in a 1-L volumetric flask, add 800 mL distilled water to dissolve and dilute to volume.
2. Standard solution containing 50 µg Al/mL, see 11.1.2.

Extraction of Al
Weigh 5 g of soil in a 100-mL polyethylene centrifuge tube, add 50 mL KCl solution, and shake for 1 h with a wristaction shaker. Centrifuge the mixture at 2400 rpm for 15 min, filter the supernatant into a 100-mL volumetric flask, and dilute to volume with distilled water.

Analysis of Extractable Al by Atomic Absorption Spectroscopy
Follow the procedure discussed above in section 11.1.2.

11.2 DETERMINATION OF IRON IN SOILS

11.2.1 Forms of Iron in Soils

Iron, Fe, is another important element in soils. It is the third most abundant element in rocks and minerals. The central core of the earth is made up mostly of iron. Plants and animals contain iron.

The most important iron mineral in soils is *hematite*, Fe_2O_3, and *magnetite*, Fe_3O_4. The hydrated form of hematite is often called *limonite*, $2Fe_2O_3 \cdot 3H_2O$. Hematite is red in color and its presence gives to the soils this red color. Magnetite is black in color and crystalline in nature. It has strong magnetic properties. Another iron mineral is *pyrite*, FeS_2, which occurs in soils as yellow crystals with a metallic luster similar to gold, hence the name *fool's gold*. Ilmenite, $FeTiO_3$, is another important iron mineral, and is a valuable source for the production of many industrial products. As is the case with Al, iron can also occur in soils as amorphous compounds, as coatings around soil particles, and as organic complexes. The granular structures in oxisols are believed to be stabilized by iron coatings around the structural units.

The total iron content in igneous, sedimentary, and metamorphic rocks is on the average 5.9% Fe. In soils, the total iron content is very variable, and can range from 10% to 90% Fe (Krauskopf, 1973). Lateritic soils and laterites are known for their high iron contents. The total iron content in these soils may range from 20% to 90% Fe (Mohr and Van Baren, 1960).

When released in soil water, iron exists in two different ionic forms, Fe^{2+} and Fe^{3+}. In anaerobic conditions, Fe^{2+} or ferrous ion is the dominant species, whereas in aerobic conditions Fe^{3+} or ferric iron is the major species. Ferric iron behaves similarly as Al^{3+} ions and will be quickly surrounded by 6 molecules of H_2O in octahedral coordination, yielding $Fe(H_2O)_6^{3+}$. The concentration of Fe^{2+} and/or Fe^{3+} in the soil solution is very small, due to the highly insoluble nature of Fe_2O_3 minerals. The concentration of soluble Fe in river water and ground water is estimated to range from 0.1 to 10 mg/L (Krauskopf, 1973; Tan, 1994).

Iron is an essential micronutrient for plant growth. It is needed for chlorophyll formation. In animals, iron is present in the blood hemoglobin, which acts as a carrier of oxygen. The normal iron concentration in plant tissue varies considerably with plant species from 100 µg/g dry matter of grasses to 1000 µg/g in alfalfa dry matter. Iron concentrations <20 µg/g in plant tissue are considered deficient, and may result in development of chlorosis, frequently manifested as yellow stripes on young leaves. Iron deficiency in soils can be induced by overliming, and the presence of high amounts of phosphates. Iron deficiency often occurs with plants growing in alkaline or calcareous soils. Therefore, iron deficiency in the United States is most likely a problem in the soils of arid and semiarid regions.

In view of the above, four types of determination of Fe will be presented: (1) Determination of total Fe, (2) determination of exchangeable Fe, (3) determination of available Fe, and (4) determination of free and amorphous Fe.

11.2.2 Determination of Total Fe

Procedures

Fluoro-Boric Acid Digestion Procedure
See sections 10.2.2 for reagents and extraction procedure.

Analysis of Total Fe by Atomic Absorption Spectroscopy
Calibrate the atomic absorption spectrophotometer, provided with a hollow cathode Fe tube, with a standard solution containing 5 µg Fe/mL. The use of this standard should give a reading of 0.18 absorbance units with the Perkin-Elmer AA instrument model 503. The standard is prepared by pipetting 5 mL of a stock solution containing 1000 µg Fe/mL (Fisher Sci.Co., Atlanta, GA) into a 1-L volumetric flask and dilute to volume with distilled water.

Aspirate the fluoro-boric acid digest and read the absorbance. If your reading is >0.18 absorbance, dilute your digest so that the reading is below or equals 0.18 absorbance. The working range for Fe is linear from 0 to 5 µg Fe/mL.

11.2.3 Determination of Exchangeable Fe

Procedure

NH_4OAc method

See sections 9.3.1 and 10.3.3 for reagents and extraction procedures.

Determination of Microelements

Analysis of Extractable Fe by Atomic Absorption Spectroscopy
Follow the procedure discussed above for analysis of total Fe by atomic absorption spectroscopy in section 11.2.2.

11.2.4 Determination of Available Fe

Available Fe, the Fe fraction in soils available to plants, can be determined by the NH_4OAc method discussed above in section 11.2.3. By definition exchangeable ions are available to plants. However, recently the use of chelating agents have been found more useful in the extraction of available Fe in soils. Several chelating agents, including EDTA, EDDHA, and DTPA, have been used for this purpose. Lindsay and Norvell (1978) have developed the DTPA (diethylenetriaminepentaacetic acid) method and reported the following correlations between DTPA-extractable Fe concentrations and plant growth:

DTPA extractable Fe	Availability index
<2.5 mg/L	Deficient
2.5 - 4.5 mg/L	Hidden hunger
>4.5 mg/L	Sufficient

Because of the above, the DTPA method will be discussed below for completeness. As is the case with the NH_4OAc method, the DTPA extract is also useful for determination of other available micronutrients, such as Cu, Zn, and Mn.

Procedure

DTPA method

Reagents
1. DTPA, diethylenetriaminepentaacetic acid, solution: Weigh

19.67 g DTPA, add 1 L of distilled water, and dissolve under constant stirring. Weigh separately 149.2 g TEA (triethanol amine), and 14.7 g $CaCl_2.2H_2O$ and add 1 L of distilled water to dissolve. Pour under constant stirring the DTPA solution into the TEA-$CaCl_2$ mixture. After the DTPA has dissolved completely, dilute the solution to 9 L. Adjust the pH to 7.3 with HCl, and make up the volume to 10 L with distilled water.
2. Standard Fe solution: Pipet 5 mL of a stock solution containing 1000 mg Fe/mL (Fisher Sci.Co., Atlanta, GA) into a 1-L volumetric flask and dilute to volume with distilled water containing 1.0 g DTPA/L. This standard solution can be used for Fe determination either by atomic absorption spectroscopy or by colorimetry. Since the atomic absorption method is currently believed a faster and accurate method than the colorimetric method, the AA method will be discussed below.

Extraction of Available Fe

Weigh 5 g of air-dried soil (<2 mm) in a 100-mL polyethylene centrifuge tube, add 20 mL DTPA mixture, and shake for 3 h with a wristaction shaker. Centrifuge at 2400 rpm, and filter the supernatant into a 50-mL volumetric flask. Dilute to volume with distilled water.

Analysis of Available Fe by Atomic Absorption Spectroscopy

Calibrate the AA instrument with the Fe standard containing DTPA, and proceed as discussed in section 11.2.2.

11.2.5 Determination of Free and Amorphous Fe

Procedures

Sodium Citrate-Dithionite-Bicarbonate Method (Jackson, 1960)

Determination of Microelements

Reagents
1. Sodium citrate, $Na_3C_6H_5O_7 \cdot 2H_2O$, 0.3 M: Weigh 88.0 g of Na-citrate in a 1-L volumetric flask, add 800 mL distilled water to dissolve under constant stirring, and dilute to volume.
2. Sodium bicarbonate, $NaHCO_3$, 1 M: Weigh 84 g Na-bicarbonate into a 1-L volumetric flask, add 800 mL water and dissolve under constant stirring. Make up to volume with distilled water.
3. Sodium chloride, NaCl, saturated solution.

Extraction of Free and Amorphous Fe
Weigh 10 g of soils (5 g for soils high in Fe) in a 100-mL polyethylene centrifuge tube. Add 40 mL Na-citrate and 5 mL $NaHCO_3$ solution, mixed, and heat the mixture to 80°C in a waterbath. After the temperature reaches 80°C, add 1 g solid $Na_2S_2O_4$, and continue heating at 80°C for another 15 min. Do not heat above 80°C. Add 10 mL of a NaCl saturated solution, cool, and centrifuge at 2400 rpm for 30 min. The clear supernatant is filtered and collected into a 100-mL volumetric flask, and diluted to volume.

Determination of Free and Amorphous Fe by Atomic Absorption Spectroscopy
Follow the procedure above in section 11.2.2.

11.3 DETERMINATION OF SILICON IN SOILS

11.3.1 Forms of Silicon in Soils
Silicon is perhaps next to oxygen in abundance in the earth's crust. In silicon chemistry, the term *silica* is used for Si compounds, whereas the term silicon is reserved for the element Si. Silica occurs in the soil as six distinct crystalline minerals: quartz, tridymite, crystobalite, coesite, strishovite, and opal. Today they are classified in the SiO_2 group of *tectosilicates*. The

minerals quartz, tridymite, crystobalite, coesite, and strishovite are all characterized by the formula SiO_2, but differ from each other in their framework of geometric arrangements of silica tetrahedrons. Opal, on the other hand, is characterized by a formula of $SiO_2 \cdot nH_2O$, and is biogenic of origin. Flint is also a silica mineral, and a great number of amorphous or noncrystalline and paracrystalline species are present in soils.

Quartz is a common mineral in acid igneous rocks, such as granite and rhyolite. It is perhaps the most common SiO_2 mineral in soils. The silicon in quartz is surrounded by 4 oxygen atoms in tetrahedral coordination. These silica tetrahedrons are linked together by sharing oxygen atoms. The Si-O-Si bonds, called the *siloxane bond*, are the strongest bonds in nature. Quartz is, therefore, very resistant to weathering, hence, very difficult to dissolve (Tan, 1994). Most of the soluble silica originate from amorphous silica and from primary minerals, such as anorthite ($CaAl_2Si_2O_8$) and albite ($NaAlSi_3O_8$). Soluble silica can also be introduced to the soil through artificial compounds. Silicates are used in detergents and as anticorrosive agents in antifreeze. The total concentration of silica in rocks is used for the distinction of igneous rocks into acid, intermediate, and basic rocks:

Igneous rocks	SiO_2 %
Acid	>65
Intermediate	65-50
Basic	<50

The silica content in soils are, of course, lower than that in the rocks above, and varies considerably depending on differences in soil formation and climate. The SiO_2 content in laterites, humid region soils affected by leaching of silica, ranges from 7 to 40% (Mohr and Van Baren, 1960). It is somewhat higher in

Determination of Microelements

spodosols, humid region soils affected by an accumulation of considerable amounts of silica. In contrast to the total SiO_2 content, the concentration of dissolved silica in natural waters is very small, ranging from 1 - 30 mg SiO_2/L, though concentrations of 100 mg SiO_2/L have sometimes been reported (McKeague and Cline, 1963; Tan, 1994). Seawater is usually low in silica content, because Si is used by marine animals for formation of their skeletons and shells. Present in amounts below 140 mg SiO_2/L, soluble silica is present mainly as *monosilicic acid*, also known as *orthosilicic acid*, which is a true solution. Several authors argue that the concentration of soluble silica is in the range of 100 to 400 mg/L, and the value of 120 mg SiO_2/L is reported as the equilibrium solubility of silica. The chemical formula of monosilicic acid can be written as H_4SiO_4, or $Si(OH)_4$, hence its ion can then be represented as either SiO_4^{4-}, or Si^{4+}, or $SiO_2(OH)_2^{2-}$. Monosilicic acid forms *silica gel* by dehydration. The dehydration process, assumed to take place stepwise through formation of meta-, di-, tri-silicic acid, etc., is a type of polymerization process.

Silicon is not considered a plant nutrient, though all plants contain Si, especially grasses. It is believed that Si in grasses increases the strength of the leafblades and straw. Rice plants are reported to be very sensitive to Si deficiency (Uchiyama and Onikura, 1955). Some of the Si compounds above are also soluble in weak alkali solutions, Na-dithionite, Na-carbonate, and NH_4OAc reagents. For analytical purposes, this type of silica is sometimes called *extractable Si* (Hallmark et al., 1982). However, its importance in plant nutrition, soil genesis and characterization is subject to many arguments. Data are unavailable indicating the presence of a correlation between extractable Si and plant growth, whereas the use of extractable Si content is uncommon in soil genesis and characterization.

In view of the above, only the determination of total Si will be discussed below. If extractable Si has to be determined, the NH_4OAc method for extraction of exchangeable cations can be employed, and the Si determined by atomic absorption spectroscopy or colorimetry as discussed below. Determination of water-soluble Si can also be performed by analyzing a water-extract by either of the methods above. However, water-soluble Si concentration is reported to increase by prolonged contact of soil with water during extraction, and to decrease with increasing pH and decreasing temperature (McKeague and Cline, 1963).

11.3.2 Determination of Total Si

Procedure

Fluoro-Boric Acid Digestion Procedure
See sections 10.2.2 and 10.3.2 for reagents and extraction procedure.

The Si in the digest can be determined either by atomic absorption spectroscopy, or by the (blue silico-molybdous acid) colorimetric method. Although both methods are suitable, Si determination by atomic absorption spectroscopy is considered a faster method. Both methods are presented below.

1. *Analysis of Total Si by Atomic Absorption Spectroscopy*
Calibrate the atomic absorption spectrophotometer, provided with a hollow cathode Si tube, with a standard solution containing 50 µg Si/mL. The use of this standard should give a reading of 0.12 absorbance units with the Perkin-Elmer AA instrument model 503. The standard is prepared by pipetting 5 mL of a stock solution containing 1000 µg Si/mL (Fisher Sci. Co., Atlanta, GA) into a 100-mL volumetric flask and diluting to volume with distilled water.

Aspirate the fluoro-boric acid digest and read the absorbance. If your reading is >0.12 absorbance, dilute your digest so that the reading is below or equals 0.12 absorbance. The working

Determination of Microelements

range for Si is linear from 0 to 50 µg Si/mL in aqueous solution.

2. *Colorimetric Determination of Total Si*
This method is provided as an alternative method to the analysis of Si by atomic absorption spectroscopy.

Blue Silicomolybdous Acid Method

Reagents
1. Ammonium molybdate solution: Weigh in a 1-L beaker 54 g of $(NH_4)_6Mo_7O_{24} \cdot 4H_2O$, add 800 mL distilled water to dissolve under constant stirring, and adjust the pH to 7.0 with NaOH. Transfer the solution into a 1-L volumetric flask, and dilute to volume with distilled water.
2. H_2SO_4 solution, 1 N
3. Tartaric acid solution. Weigh 100 g tartaric acid in a 500-mL volumetric flask, add 400 mL distilled water to dissolve under constant stirring, and dilute to volume.
4. Reducing agent: Weigh 25 g sodium bisulfite, $NaHSO_3$, in a 250-mL volumetric flask. Weigh separately 2 g Na_2SO_3 and 400.0 mg of 1-amino-2-naphthol-4-sulfonic acid in a beaker and dissolve them in 25 mL water. Add this mixture to the sodium bisulfite solution, and dilute to volume with distilled water. Store the reducing agent in a polyethylene bottle.
5. Standard Si solution (50 mg Si/L). Pipet 5 mL of a stock solution containing 1000 µg Si/mL into a 100-mL volumetric flask, and dilute to volume with distilled water.

Analysis of Total Si
Pipet 5 mL of the fluoro-boric acid digest into a 50-mL volumetric flask, add 10 mL H_2SO_4 and 10 mL of ammonium molybdate solution. Wait two min, after which 5 mL of tartaric acid and 1 mL of reducing agent are added. Dilute to volume with distilled water, shake well to develop the blue color, and measure the absorbance at 400 nm. Prepare a standard curve

by pipetting 0, 2, 5, and 8 mL of the standard solution (50 μg Si/mL) into a series of 50-mL volumetric flasks, and proceeding as discussed above.

11.4 DETERMINATION OF MANGANESE IN SOILS

11.4.1 Forms of Manganese in Soils

Manganese is present in small quantities in many crystalline rocks. It is released into the soil by rock weathering and is redeposited in various forms of Mn oxides, e.g., *pyrolusite*, MnO_2, *braunite*, Mn_2O_3, and *manganite*, $MnO(OH)$ or $Mn_2O_3 \cdot H_2O$. Pyrolusite is a mineral that belongs to the rutile group, which is characterized by the formula XO_2. If X=Ti, the mineral is rutile, but when X=Mn, the mineral is pyrolusite. Uranium oxide also belongs to this group. Pyrolusite is an accessory mineral in crystalline rocks. (Tan, 1994).

The total Mn content in soils varies considerably from 20 mg/kg to 6000 mg/kg (Krauskopf, 1973). The element can exist in three oxidation states: Mn^{2+}, Mn^{3+}, and Mn^{4+}. This is in contrast to iron, which exists only in two oxidation states. The divalent manganese ion is the main form of Mn in soil water, and especially in reducing soil environments. Because of its cationic nature, Mn^{2+} is usually adsorbed on the negatively charged surfaces of soil colloids. It is then called *exchangeable manganese*. The trivalent form usually exists as Mn_2O_3, which can be found in substantial amounts in acid soils. The trivalent ion itself is unstable in soil water. The tetravalent form, MnO_2, is perhaps the most stable and inert form of manganese. The oxide occurs usually in a strongly oxidized soil environment, and its ion can only exist in very strongly acid soils.

The concentration of Mn^{2+} in soil water is very low, seldom exceeding 0.05 mg/L. Its concentration increases at low pH values, and decreases at high pH values. Therefore, high amounts of manganese may be present in highly weathered acid

Determination of Microelements

soils, such as ultisols and oxisols. On the other hand, low amounts of manganese usually occur in aridisols.

Manganese is an essential micronutrient and is needed by plants to activate a number of enzymes, e.g., decarboxylase, dehydrogenase, and oxydase enzymes. It also plays an essential role in photochemistry and in N-metabolism and assimilation. Manganese is required for plant growth only in very small amounts. The normal Mn content in plant tissue is approximately 20 - 500 µg/g dry matter. A Mn concentration of < 20 µg/g dry matter is considered deficient and is usually reflected by development of necrotic spots, known as *marsh spots* in pea leaves, and *gray specks* in oat leaves. Soybean and oats are especially sensitive to Mn deficiency.

The critical Mn concentration in soils is in the range of 2-9 µg/mL for Mn extracted by the double acid procedure (Cox and Kamprath, 1973). A double acid extractable Mn concentration lower than this critical level results in Mn deficiency. On the other hand, large amounts of Mn cause Mn toxicity in plants. Mn concentrations in plant tissue of >500 µg/g dry matter are excessive, and may create toxicity. The latter can also cause necrosis and chlorosis in leaves. Mn toxicity usually occurs in acid soils, and can be controlled by liming.

In view of the different forms and their importance in soil formation, characterization and plant nutrition, the following procedures will be presented below: (1) determination of total Mn, (2) determination of exchangeable Mn, and (3) determination of available Mn.

11.4.2 Determination of Total Mn

Total elemental analysis can, of course, be done by several methods, e.g. Na_2CO_3 fusion or fluoro-boric acid digestion. For the purpose of being consistent and for reasons stated earlier, the fluoro-boric acid digestion method is presented here. The preferred analysis of Mn concentration in the digest is by atomic absorption spectroscopy. Colorimetric analysis of Mn is cumbersome and time consuming.

Procedure

Fluoro-Boric Acid Digestion Procedure
See sections 10.2.2 and 10.3.2 for reagents and extraction procedure.

Analysis of Total Mn by Atomic Absorption Spectroscopy
Calibrate the atomic absorption spectrophotometer, provided with a hollow cathode Mn tube, with a standard solution solution containing 2 µg Mn/mL. The use of this standard should give a reading of 0.16 absorbance units with the Perkin-Elmer AA instrument model 503. The standard is prepared by pipetting 2 mL of a stock solution containing 1000 µg Mn/mL (Fisher Sci. Co., Atlanta, GA) into a 100-mL volumetric flask and diluting to volume with distilled water.

Aspirate the fluoro-boric acid digest and read the absorbance. If your reading is >0.16 absorbance, dilute your digest so that the reading is below or equals 0.16 absorbance. The working range for Mn is linear from 0 to 3 µg Mn/mL in aqeous solution.

11.4.3 Determination of Exchangeable Mn

Procedures

NH_4OAc Method
See sections 9.3.1 and 10.3.3 for reagents and extraction procedures.

Analysis of Exchangeable Mn by Atomic Absorption Spectroscopy
Follow the procedure discussed above for analysis of total Mn

Determination of Microelements

by atomic absorption spectroscopy in section 11.4.2.

11.4.4 Determination of Available Mn

Available Mn includes NH_4OAc-extractable, water-soluble, DTPA-extractable, and double acid extractable Mn. Ammonium acetate extractable Mn is covered in the preceding section, whereas DTPA extraction is discussed in procedures for extraction of available Fe (section 11.2.4). The same extract can be analyzed for available Mn content. Therefore, the double acid procedure will be presented below, this method has also been discussed earlier for the extraction of soluble P (section 10.2.4).

Extraction of Available Mn by the Double Acid Procedure (Mehlich's No.1)

For reagents and procedure see section 10.2.4, determination of soluble P in dilute $HCl+H_2SO_4$ (Mehlich's No.1 method).

Analysis of Available Mn by Atomic Absorption Spectroscopy

Follow the procedure discussed above for analysis of total Mn by atomic absorption spectroscopy in section 11.4.2.

11.5 DETERMINATION OF COPPER IN SOILS

11.5.1 Forms of Copper in Soils

Copper, Cu, is classified as a *native element*, meaning that it can occur as a native element in the earth crust in contrast to such elements as Al, which exist only in the form of a compound. Copper belongs to the gold group, which includes gold, Au, silver, Ag, and copper, Cu, and is, therefore, a noble element. However, the most common form of Cu in soils is in the form of minerals, e.g., in the forms of sulfides, sulfosalts, oxides, carbonates, and silicates. Many of these minerals are major constituents of copper deposits in copper mines (Hurlbut and Klein, 1977). Most of these copper deposits produce upon weathering exotic minerals, with brilliant green and blue colors, e.g. *malachite*, $Cu_2(OH)_2CO_3$, and *azurite*, $Cu_3(OH)_2(CO_3)_2$. When polished properly, these minerals are frequently used as semi-precious jewelry. Most of these copper minerals are rare in soils. The most common in soils are perhaps chalcopyrite, $CuFeS_2$, related to pyrite or fool's gold, FeS_2, and copper oxide, Cu_2O. Copper can also be introduced in soils by Cu-containing pesticides. It is sometimes added to poultry and pig feed as a growth rate regulator, and using their manure as soil amendment increases Cu content in soils.

The total Cu content in soils is in the range of 10 mg/kg to 80 mg/kg (Krauskopf, 1973). In the soil solution, copper exists as Cu^+ (cuprous), or Cu^{2+} (cupric) ions. At concentrations >10^{-7} mol/L Cu^+ ions are unstable at ordinary temperatures. However, it can exist at these conditions as a $CuCl_2^-$ chelate (Krauskopf, 1973). Cuprous ions are so unstable in aqueous solution that they will be automatically affected by the soil redox process, called *auto-reduction-oxidation* reaction, and converted into Cu and Cu^{2+}:

$$2Cu^+ \rightarrow Cu + Cu^{2+}$$

Therefore, cupric ions are more stable and are the major or common copper ions in the soil solution. However, Lindsay (1973) believes that Cu^{2+} is the dominant ion in soils only at pH values <7.3, whereas above this pH $Cu(OH)^+$ is the major copper ion. The concentration of soluble Cu ions in soils are on the average 20 mg/L. Because of its cationic nature, Cu^{2+} will be adsorbed by negatively charged surfaces of soil colloids, and are then called exchangeable Cu. The concentration of free and exchangeable Cu ions are high in acid soils, and low in basic soils. The Cu^{2+} concentrations in many neutral and calcareous

Determination of Microelements

soils from Colorado are reported to amount only 1×10^{-12} *M* (Norvell, 1973). Therefore, Cu deficiency is more likely to occur in basic soils, e.g. aridisols, whereas Cu toxicity may be displayed more by highly weathered acid soils, e.g. ultisols and oxisols.

Copper is a micronutrient, and is needed only in small amounts by plants. The element is required for chlorophyll formation, hence Cu will affect photosynthesis. It is essential in enzyme reactions, and in the reproductive stages. It is present in lactase and several other enzymes. Its presence in the plant cytochrome oxidase has created speculation for its critical role in metabolism. Cu is noted to be needed in protein and carbohydrate metabolism, and in nitrogen fixation. The normal concentration in plant tissue is approximately 4 - 30 µg Cu/g. Present at concentrations <3 µg/g dry matter, Cu deficiency will occur, which is manifested by yellowing and curling of leaves, stunted growth, and development of short internodes. On the other hand, Cu concentrations in leaf tissue >20 µg/g will result in Cu toxicity. Leaves are yellow, roots are poorly developed and frequently discolored.

From the above, it can be noticed that Cu is not used as an important parameter in soil characterization, soil genesis and classification. Its importance is only in soil fertility and plant nutrition. Hence, only the following three procedures for determination of Cu will be discussed in this book: (1) determination of total Cu, (2) determination of exchangeable Cu, and (3) determination of available Cu. It must be noticed, however, that exchangeable Cu is also available to plant growth.

11.5.2 Determination of Total Cu

Procedure

Fluoro-Boric Acid Digestion Procedure

See sections 10.2.2 and 10.3.2 for reagents and extraction procedure.

Analysis of Total Cu by Atomic Absorption Spectroscopy

Calibrate the atomic absorption spectrophotometer, provided with a hollow cathode Cu tube, with a standard solution containing 5 µg Cu/mL. The use of this standard should give a reading of 0.25 absorbance units with the Perkin-Elmer AA instrument model 503. The standard is prepared by pipetting 5 mL of a stock solution containing 1000 µg Cu/mL (Fisher Sci.Co., Atlanta, GA) into a 1-L volumetric flask and diluting to volume with distilled water.

Aspirate the fluoro-boric acid digest and read the absorbance. If your reading is >0.25 absorbance, dilute your digest so that the reading is below or equals 0.25 absorbance. The working range for Cu is linear from 0 to 5 µg Cu/mL in aqeous solution.

11.5.3 Determination of Exchangeable Cu

Procedures

Two methods can be used for the determination of exchangeable Cu, e.g., NH_4-acetate method, and Mehlich's No.1 method. Though both methods are equally suitable, some authors prefer the NH_4-acetate method, whereas others prefer the double acid procedure, also called Mehlich's No.1 method. Therefore, the two methods are presented below.

1. *NH_4OAc Method*

See sections 9.3.1 and 10.3.3 for reagents and extraction procedures.

Determination of Microelements

2. *Extraction of Exchangeable Cu by the Double Acid Procedure (Mehlich's No.1)*
For reagents and procedure see section 10.2.4, determination of soluble P in dilute $HCl+H_2SO_4$ (Mehlich's No.1 method).

Analysis of Exchangeable Cu by Atomic Absorption Spectroscopy
See analysis of total Cu by atomic absorption spectroscopy.

11.5.4 Determination of Available Cu

DTPA Method
See section 11.2.4 for reagents and extraction procedures.

Analysis of Available Cu by Atomic Absorption Spectroscopy
See analysis of total Cu by atomic absorption spectroscopy.

11.6 DETERMINATION OF BORON IN SOILS

11.6.1 Forms of Boron in Soils

Boron is a non-metal, in contrast to the other micronutrient elements. It is present in small amounts in igneous, sedimentary and metamorphic rocks. In areas with geothermal activity, gases flowing from the inner of the earth contains B. When dissolved in hot springs the element becomes boric acid, H_3BO_3, which is acclaimed for its medicinal properties. However, the major inorganic sources of B are borate and boron-silicate minerals. Borate minerals occurs mostly in arid regions. They are formed by evaporation of water in enclosed salt water lakes and basins in arid regions. The BO_3 units polymerize and forms a great variety of borate crystals. The most common borate minerals are (1) *borax*, $Na_2B_4O_5(OH)_4 \cdot 8H_2O$, (2) colema-

nite, $CaB_3O_4(OH)_3 \cdot H_2O$, (3) *kernite*, $Na_2B_4O_6(OH)_2 \cdot 3H_2O$, and (4) *ulexite*, $NaCaB_5O_6(OH)_6 \cdot 5H_2$). Borax is the most widespread of the four. In contrast, boron silicate is not confined to arid regions. They are cyclosilicates and have been formed by cooling and crystallization of molten silicate materials, such as lava and magma. Two of the major boro-silicate minerals are *tourmaline* and *axenite*. Both occurs as mineral inclusions in granite rocks. Tourmaline is considered one of the most beautiful semi-precious gemstone, and is also referred to as *Brazilian emerald*.

The total B content in soils is approximately between 7 and 80 µg/mg. It is released by weathering of the minerals in the form of H_3BO_3 and its concentration is usually $< 1 \times 10^{-4}$ M, which equals 0.1 mg/L (Krauskopf, 1973), a very small amount. Concentrations of 10 mg B/mL are perhaps more representative for surface soils. In the soil solution, B can exist both as H_3BO_3 and $H_2BO_3^-$. High concentrations of B is only found in seawater, where an average concentration of 4.6 µg/mL has been reported (Krauskopf, 1973). Because of its anionic character, BO_3^{2-} will not be attracted by the negatively charged soil colloids. In humid regions borate ions may tend to leach from soils. Therefore, soluble B concentrations are low in highly leached soils. Since highly leached soils usually exhibit low soil pH, acid soils are deficient in B. On the other hand, soluble B concentrations are expected to be high in soils not affected by leaching. Since arid region soils are usually not affected by leaching, and are, therefore, strongly basic in reaction, they may contain excessive amounts of B for plant growth. Therefore, B deficiency is more likely to occur in ultisols and oxisols, whereas B toxicity will likely be noted in aridisols. This is in contrast with the other micronutrient elements.

Boron is an essential micronutrient element for plant growth. It is needed for cell division, hence for the growth of young shoots. It is essential in sugar translocation and in the synthesis of hormones and protein in plants. Germination of

Determination of Microelements

pollen grains, carbohydrate metabolism, and activity of dehydrogenase enzymes are influenced by B. The normal B concentration in plant tissue is reported to be between 20 and 100 µg/mg dry matter of mature leaves. A B content of <15 µg/mg indicates B deficiency, whereas a B concentration >200 µg/mg indicates B toxicity. Boron deficiency is usually manifested in formation of white and rolled leaves. In sugar beets, it results in rotting of the shoots, a nutritional disease called *heart rot*.

Because B in soils is primarily of importance in soil fertility and plant nutrition and is not used as a significant parameter in soil characterization, soil genesis and classification, only two types of B determination will be presented in this book: (1) determination of total B, (2) determination of available B.

11.6.2 Determination of Total B

Since B is to be determined, the fluoro-boric acid digestion procedure can not be used for this analysis. Instead, the Na_2CO_3 fusion method is selected and discussed below.

Procedure

Reagents
1. Na_2CO_3, anhydrous, solid crystal powder.
2. Na_2CO_3, 30%: Weigh 30 g of Na_2CO_3 in a 100-mL volumetric flask, add 80 mL of distilled water, and dilute to volume. Store in a polyethylene vial.
3. H_2SO_4, 4 N: Measure 112 mL of conc H_2SO_4 into a 1-L volumetric flask, add 500 mL distilled water and allow the solution to cool. Dilute to volume with distilled water.
4. HCl, 0.1 N: Measure 8.1 mL conc HCl into a 1-L volumetric flask, add 500 mL distilled water and allow the solution to cool. Dilute to volume with distilled water.
5. Ethyl alcohol, 95%

6. NaOH, 0.01 N: Weigh 40 mg NaOH in a 100-mL volumetric flask, add 80 ml distilled water to dissolve under constant stirring, and dilute to volume.
6. Phenolphthalein solution: Weigh 50.0 mg phenolphthalein in a 100-mL volumetric flask, add 50 mL ethyl alcohol, and dilute to volume with distilled water. Shake the flask until the phenolphthalein crystals have dissolved completely.
7. Standard B solution: Pipet 35 mL of a stock solution containing 1000 µg B/mL (Fisher Sci.Co., Atlanta, GA) into a 100-mL volumetric flask, and dilute to volume with ethyl alcohol. This solution, called standard solution, contains 350 µg B per mL.

Na_2CO_3 Fusion

Weigh 1.0 g of soil, ground to pass a 100-mesh sieve, in a Pt crucible, add 6 g of Na_2CO_3 powder, and heat with a bunsen burner until sample is completely fused with the Na_2CO_3. Cool, add 10 mL of H_2SO_4 to disintegrate and dissolve the melt, and place the crucible on its side in a 250-mL Pyrex beaker. Continue adding 4 N H_2SO_4 in 4-mL increments until the solution has a pH of 6.0 -6.5. Check pH with a pH meter. Filter the solution into a 500-mL volumetric flask, and wash the contents of the beaker and the crucible by spraying with water into the filter, keeping the volume to approximately 150 ml.
Dilute with ethyl alcohol to a volume of 400 mL, and add a few drops of phenolphthalein. Add Na_2CO_3 (30%) solution to attain a slightly alkaline reaction, and dilute to volume with ethyl alcohol. Prepare a blank by following the same fusing procedure above without soil sample.

Analysis of Total B by Atomic Absorption Spectroscopy

Calibrate the atomic absorption spectrophotometer, provided with hollow cathode B tube, with a standard solution containing 350 µg B/mL. This standard solution will give a reading of 0.1 absorbance unit with a Perkin-Elmer AA instrument, model

Determination of Microelements 215

503. The working range for B is linear for concentrations between 0 - 500 µg B/mL.

11.6.3 Determination of Available B

Procedure

Hot water extract method

Reagents
1. $Ca(OH)_2$ suspension: Weigh 400.0 mg $Ca(OH)_2$ into a 100-mL volumetric flask, and dilute to volume with distilled water.
2. HCl, 0.1 N: Measure 8.1 mL conc HCl into a 1-L volumetric flask, add 800 mL distilled water, and allow the solution to cool. Dilute to volume with water.
3. $CaCl_2$, 0.01 M: Weigh 1.11 g anhydrous $CaCl_2$ into a 1-L volumetric flask, add 900 mL distilled water to dissolve under constant stirring, and dilute to volume with distilled water.

Hot Water Extraction of Available B
Weigh 20 g of airdry soil into a 250-mL Erlenmeyer flask, and add 40 mL of the $CaCl_2$ solution. Place a refluxing funnel on the Erlenmeyer, and heat the mixture until boiling, and reflux the suspension for 5 min. Cool and filter the suspension into a 50-mL volumetric flask, and dilute to volume with distilled water. Pipet 25 mL of the filtrate into an evaporating dish, add 2 mL of the $Ca(OH)_2$ suspension, and evaporate to dryness on a waterbath. Heat the evaporating dish gently over a flame to destroy organic matter. Cool to room temperature, add 5 mL of HCl (0.1 N), and dissolve the digest by rubbing using a rubber policeman. Filter the digest into a 10-mL volumetric flask, and dilute to volume with distilled water.

Analysis of Available B by Atomic Absorption Spectroscopy
See analysis of total B by atomic absorption spectroscopy (section 11.6.2).

11.7 DETERMINATION OF MOLYBDENUM IN SOILS

11.7.1 Forms of Molybdenum in Soils

Molybdenum is a rare element in soils, and is present only in very smalll amounts in igneous and sedimentary rocks. The major inorganic source of Mo is *molybdenite*, MoS_2, an important molybdenum ore. Molybdenite is an accessory mineral in granite rocks. The mineral can be changed by weathering into a yellow colored *ferrimolybdate* mineral, $Fe_2(MoO_4)_3 \cdot 8H_2O$, or oxidized into yellow-white molybdenum oxide, MoO_3. Other Mo minerals are very rare, e.g. *wulfenite*, $PbMoS_4$, and *powellite*, $CaMoO_4$. Wulfenite is present in the oxidized fraction of lead veins, and is considered a minor source of Mo.

The total Mo content in soils is perhaps the lowest of all the micronutrient elements, and is reported to range between 0.2 µg/g and 10 µg/g. According to Brady (1990) the average Mo content in surface soils is approximately 2 µg/g, but Kubota and Cary (1982) reported that the average Mo content in U.S. soils was 1.25 µg/g. In the soil solution Mo can exists in the form of MoO_4^{2-} and $HMoO_4^-$ ions. In alkaline solutions, MoO_4^{2-} is the dominant ion, whereas at pH 5-6, $HMnO_4^-$ ions are dominant. At very low pH values Mo is present as H_2MoO_4 and MoO_2^{2+} (Krauskopf, 1973). However, because of the extremely low pH, the undissociated and cationic forms of Mo are not important in soils. Consequently only the anionic form, MoO_4^{2-}, is of significance in agricultural soils. The other anion, $HMoO_4^-$, will play a role only in soils with high pH, e.g. aridisols. Because of its anionic nature, MoO_4^{2-} ions will not be attracted by negatively charged colloids, and tend to be leached in humid region soils. Therefore, soluble Mo concentration is more likely to be very low in acid soils, and high in basic soils. In fact, Mo-oxide, MoO_3, will dissolve in alkaline condition to produce molybdates, increasing in this way the soluble Mo content. Accordingly, Mo toxicity will occur more in alkaline soils of the arid regions, whereas Mo deficiency will be found more in acid soils of the humid regions. This is in contrast with the other micronutrient element. As discussed earlier with B toxicity, borate and molybdate anions create toxicity in alkaline soils, whereas Al, Fe, Cu, Zn, and Mn ions may induce toxicity in acid soils.

Mo is an essential micronutrient elements for plant growth. It is needed in nitrogen fixation, since it is present in nitrogenase, and in nitrate reductase enzymes. Mo is also required in nitrogen assimilation. The Mo concentration in plant tissue varies from 0.1 - 300 µg/g dry matter. A Mo content of <0.1 µg/g dry matter may cause Mo deficiency, which is reflected by mottling between leaf-veins, and formation of necrotic spots at leaf margins, and tips. Since Mo is essential in N-fixation, legume plants can show Mo deficiency symptoms, when the rhizobia bacteria in the root nodules are deficient in Mo for their N fixation reaction. On the other hand, a Mo content of 15 to 20 µg/g dry matter or higher in forage crops may cause toxicity to cattle, although these concentrations may not be harmful to the crops themselves. This physiological sickness, called *molybdenosis*, has been observed in the western part of the U.S., and in the Everglades in Florida (Kubota and Cary, 1982).

As is the case with B, Mo content is not a significant parameter in soil characterization, soil genesis and classification. It is important only in soil fertility, plant nutrition, and environmental problems. Therefore, the following determinations of Mo will be discussed in this book: (1) determination of total Mo, and (2) determination of available Mo.

11.7.2 Determination of Total Mo

Procedure

Fluoro-Boric Acid Digestion Procedure
See sections 10.2.2 and 10.3.2 for reagents and extraction procedure.

Analysis of Total Mo by Atomic Absorption Spectroscopy
Calibrate the atomic absorption spectrophotometer, provided with a hollow cathode Mo tube, with a standard solution containing 40 µg Mo/mL. The use of this standard should give a reading of 0.35 absorbance units with the Perkin-Elmer AA instrument model 503. The standard is prepared by pipetting 4 mL of a stock solution containing 1000 µg Mo/mL (Fisher Sci. Co., Atlanta, GA) into a 100-mL volumetric flask and diluting to volume with distilled water.

Aspirate the fluoro-boric acid digest and read the absorbance. If your reading is >0.35 absorbance, dilute your digest so that the reading is below or equals 0.35 absorbance. The working range for Mo is linear from 0 to 40 µg Mo/mL in aqueous solution.

11.7.3 Determination of Available Mo

Procedures

NH_4OAc Method
See sections 9.3.1 and 10.3.3 for reagents and extraction procedures.

Analysis of Available Mo by Atomic Absorption Spectroscopy
See section 11.7.2, analysis of total Mo by atomic absorption spectroscopy.

11.8 DETERMINATION OF ZINC IN SOILS

11.8.1 Forms of Zinc in Soils

Zn is present in small quantities in igneous and sedimentary rocks. Zn concentrations are usually higher in basic igneous rocks, such as basalt, than in acid igneous rocks, such as granite. The major inorganic sources of Zn are *sphalerite*, ZnS, *smithsonite*, $ZnCO_3$, and *hemimorphite*, $Zn_4(Si_2O_7)(OH)_2 \cdot H_2O$. *Hydrozincite*, $Zn_5(CO_3)_2(OH)_6$, is another Zn mineral frequently mentioned, but this mineral is considered similar as smithsonite (Hurlbut and Klein, 1977). All the minerals stated above are very rare in soils, and occur only in large amounts as Zn ore deposits. Sphalerite is perhaps the most important and common ore in Zn mines. Smithsonite is a Zn-carbonate and belongs to the calcite group. It is usually found with Zn deposits in limestones. The Zn-silicate hemimorphite is a *sorosilicate* and is classified as a secondary mineral, since it is formed by oxidation of Zn deposits containing sphalerite and smithsonite.

The Zn metal itself is rapidly oxidized in moist air and converted into a basic Zn-carbonate, $Zn_2CO_3(OH)_2$, which coats the metal and prevents it from further corrosion. The concentration of total Zn in soils is approximately in the range of 10 µg to 300 µg/g. This concentration may perhaps be higher in soils in the vicinity of Zn mines and Zn smelters. The average Zn content in normal agricultural soils is 50 µg/g (Brady, 1990). In the soil solution Zn is present as Zn^{2+} ions, which are the dominant ions at pH <9.0. Most Zn compounds are soluble in water. Because of its cationic nature, the Zn ion is adsorbed by the negatively charged surfaces of soil colloids, and are then called *exchangeable Zn*. The concentration of Zn ions are dependent upon pH, as is the case with all the other microelements. At low pH, Zn^{2+} concentration is very high in soils, whereas at high pH, Zn^{2+} concentration is low. At pH 9.5, Zn^{2+} exhibits minimum solubility and precipitates as $Zn(OH)_2$. Therefore, high soluble Zn contents occur in acid soils, whereas low soluble Zn contents are found more in basic soils, especially in aridisols. For example, the concentration of Zn^{2+} ions in neutral and calcareous soils from Colorado is in the range of 1 x 10^{-8} to 1 x 10^{-10} M (Norvell, 1973). Consequently, Zn toxicity is more likely to occur in acid soils, whereas Zn deficiency will be found more in soils with very strongly basic reaction.

Zinc is a micronutrient to plants, and is therefore needed only in very small amounts. The nutrient element functions as a catalyst. It is present in several plant enzymes, e.g., dehydrogenase, proteinase, and peptidase. Zinc is also essential for seed and grain production, and development of growth hormones. Plants deficient in Zn have been noted to be also deficient in auxins (Skoog, 1940).

The normal concentration of Zn in most plants ranges on the average from 15 to 125 µg/g. Zn concentrations < 15 µg/g of leaf dry matter cause Zn deficiency, which is manifested in stunted growth. In corn plants, Zn deficiency produces white to yellow leaves with bleached stripes, a symptom known as *white bud* of corn. On the other hand, Zn concentrations >400 µg/g of leaf dry matter are considered excessive and may induce Zn toxicity.

Since the importance of Zn in soils is primarily in soil fertility and plant nutrition, as discussed above, only three types of Zn determination will be presented in this book: (1) determination of total Zn, (2) determination of exchangeable Zn, and (3) determination of available Zn.

11.8.2 Determination of Total Zn

Procedure

Fluoro-Boric Acid Digestion Procedure

See sections 10.2.2 and 10.3.2 for reagents and extraction procedure.

Determination of Microelements

Analysis of Total Zn by Atomic Absorption Spectroscopy

Calibrate the atomic absorption spectrophotometer, provided with a hollow cathode Zn tube, with a standard solution containing 0.5 µg Zn/mL. The use of this standard should give a reading of 0.12 absorbance units with the Perkin-Elmer AA instrument model 503. The standard is prepared by pipetting 0.5 mL of a stock solution containing 1000 µg Zn/mL (Fisher Sci. Co.) into a 1-L volumetric flask and diluting to volume with distilled water.

Aspirate the fluoro-boric acid digest and read the absorbance. If your reading is >0.12 absorbance, dilute your digest so that the reading is below or equals 0.12 absorbance. The working range for Zn is linear from 0 to 1 µg Cu/mL in aqeous solution.

11.8.3 Determination of Exchangeable Zn

Procedures

Two methods can be used for determination of exchangeable Zn, e.g., NH_4-acetate method, and Mehlich's No.1 method. Though both methods are equally good, some authors prefer the NH_4-acetate method, whereas others like more the double acid method developed by Mehlich. The two methods are presented below.

1. *NH_4OAc Method*

See sections 9.3.1 and 10.3.3 for reagents and extraction procedures.

2. *Extraction of Exchangeable Zn by the Double Acid Procedure (Mehlich's No.1)*

For reagents and procedure see section 10.2.4, determination of soluble P in dilute $HCl+H_2SO_4$ (Mehlich's No.1 method).

Analysis of Exchangeable Zn by Atomic Absorption Spectroscopy
See analysis of total Zn by atomic absorption spectroscopy.

11.8.4 Determination of Available Zn

Procedure

DTPA Method
See section 11.2.4 for reagents and extraction procedures.

Analysis of Available Zn by Atomic Absorption Spectroscopy
See analysis of total Zn by atomic absorption spectroscopy.

CHAPTER 12

DETERMINATION OF SOIL ORGANIC MATTER

12.1 DEFINITION AND PRINCIPLES

Soil organic matter is by definition the organic fraction derived from living organisms. It includes the living organisms, partly decomposed and decomposed plant and animal residue. The decomposed organic fraction is usually called "humus". It is composed of (1) nonhumic substances, and (2) humic substances. The nonhumic substances include carbohydrates, amino acids, lipids, lignins, etc., all of which are the metabolic products of organisms. On the other hand, humic substances, such as humic acids and fulvic acids, which are brown to black in color, are high molecular weight compounds that are synthesized by soil microorganisms. Approximately 50% to 85% of the total organic matter content is humus. About 65% to 75% of this humus is composed of humic matter, while the remaining 35 to 25% consists of nonhumic matter (Stevenson, 1979; Schnitzer, 1982).

This organic fraction of the soil is constantly undergoing physical and chemical changes as a result of decomposition and mineralization processes. The end result of these processes is the production of CO_2, H_2O, nutrients, and inorganic and organic acids. Organic matter contents in mineral soils of warm humid regions are relatively small. Mineral soils generally contain 20% organic matter and 80% mineral matter on a weight basis. On a volume basis, a fertile, loamy top soil has an average organic matter content of only 5%. In less fertile soils and in subsoils, the organic matter content is lower than 5% by volume. Generally, soils with comparatively higher organic matter content are considered more fertile than soils low in organic matter content. Depending on climatic conditions and management practices, the level of organic matter in soils generally stabilizes at a certain value. Under the warm and humid climate of the southern region in the U.S., soil organic matter seldom exceeds 3.5% (Beaty and Tan, 1972; Tan et al., 1975).

Notwithstanding its low content, organic matter is a very important soil constituent. It affects the physical, chemical and biological properties of soils. From the standpoint of the physical properties, organic matter increases the water-holding capacity of soils and promotes the development of stable soil structures by increasing granulation. Chemically, it is a source of plant nutrients, especially of N and S, in soils. The total N content is found to be a function of soil organic matter content. Statistical studies indicate the presence of a linear relationship between the organic C and N content. This relationship is formulated as follows:

$$N = a + bC$$

in which N = % total soil N, a = intercept with Y-axis, b = the regression coefficient, and C = % organic C (Tan, 1964; Tan and Troth, 1981). Biologically, organic matter is the source of food and energy for microorganisms. These organisms are vital to the many biochemical reactions in soils, such as ammonification, ni-

Determination of Soil Organic Matter

trification, N-fixation, and nutrient cycling.

Upon decomposition of organic matter, inorganic nutrients in the plant tissue, N, P, K, Ca, Mg, Fe, Cu, Zn and Mn, etc., are released into the soil. At the same time, humus, is formed. Humus, a byproduct, which increases the cation exchange capacity of soils, is characterized by a C/N ratio close to 10:1. The C/N ratio is by definition the ratio of % C_{org} and % N. It is often used as an index or a measurement of humification and humus quality. In the decomposition process, the C/N ratio usually decreases from a high of 80 in straw or fresh plant material to a low of 8 to 15 in humus.

A direct result of the importance of soil organic matter has been the development of many methods for its determination. However, the methods are subject to arguments, and a generally satisfactory method for soil organic matter determination has yet to be formulated. The latter is attributed to the high variability in content and composition of soil organic matter. In most cases the method used should also depends upon the objective. As previously discussed, soil organic matter is composed of all living and dead organisms, and humus. Because of this, an analysis can be based either on total organic matter content or on humus content.

In the quantitative determination of soil organic matter, it is customary to measure the C_{org} content, not the total C content of soils. The total C content of the soil is composed of C_{org} plus C_{inorg} content. The soil organic matter content is then obtained by multiplying the C_{org} concentration by a factor. This factor varies from 1.724 to 2.0 (Jackson, 1958; Nelson and Sommers, 1982).

The qualitative determination of soil organic matter is concerned mostly with the fractionation of the humic fraction, the identification of the various fractions, and the determination of their physico-chemical properties (Schnitzer, 1982). The determination of some of the chemical characteristics, e.g., total acidity, carboxyl and phenolic-OH group content, will be discussed in this section of the book. Others, such as the determination of infrared spectral characteristics, will be presented in chapter 15.

12.2 DETERMINATION OF TOTAL ORGANIC C

Total organic C can be determined by the $K_2Cr_2O_7$ wet combustion method or by the dry combustion method. The wet combustion method measures organic carbon, but one may question whether the $K_2Cr_2O_7$ used is strong enough to oxidize all the organic matter present so that the result reflects *total* C_{org} content. On the other hand, the dry combustion method determines total C content, which in some soils equals C_{org} + C_{inorg}. The inorganic carbon content may be substantial, especially in calcareous soils rich in $CaCO_3$. However, if pretreatments are applied to remove the C_{inorg} content, the dry combustion method is faster and more reliable than the wet combustion method in the determination of total C_{org} content. In acid soils, the pretreatment step can be omitted, and the soil can be used directly for the analysis. The dry combustion method, an adaptation of the Dumas or Pregl method (Pregl, 1949) is presented below.

12.2.1 Dry Combustion Method

Reagents
1. Pure oxygen gas
2. Pressurized air
3. Helium, He, gas
4. Sodium hydroxide, NaOH, pellets.
5. Manganese oxide, MnO_2, for removal of S and halogens, produced by the combustion of the sample
6. Magnesium perchlorate, $Mg(ClO_4)_2$, for removal of H_2O produced by the combustion of the sample
7. Calcium chloride, $CaCl_2$, for use in the CO_2 adsorption tube.

Determination of Soil Organic Matter

8. Sulfuric acid, H_2SO_4.

Pretreatment

This pretreatment procedure is for use with calcareous soils only. Acid soils can be analyzed directly. Weigh 500.0 mg of soil that has passed a 100 mesh sieve in a combustion boat. Digest the sample with an excess of a 5% H_2SO_4 solution for several hours. Dry the sample by placing the boat over NaOH pellets in an evacuated desiccator. Repeat this treatment until CO_2 is no longer produced when H_2SO_4 is added.

Procedure

Place the boat containing the pretreated sample in the quartz tube of the combustion unit as shown in Figure 12.1. Weigh the CO_2-adsorption tube. While the CO_2-adsorption tube is disconnected, clean the quartz tube from CO_2 and other gases by passing CO_2-free oxygen through the combustion train for at least 2 min. Connect the CO_2-adsorption tube to the combustion unit and start the analysis by heating the sample at 570°C for 10 minutes under a constant flow (5 mL/minute) of oxygen. Bring the furnace temperature up to 900°C and continue heating for another 10 minutes. The sample is then cooled in an air stream for 15 minutes. Disconnect the CO_2-adsorption tube, wipe it clean with a flannel cloth, and weigh to the nearest 0.1 mg. Run a blank in the same manner.

Calculations

g CO_2 in sample = final weight of tube - initial weight of tube

g CO_2 in blank = final weight of tube - initial weight of tube

$$\% \text{ total } C_{org} = \frac{g\ CO_{2(sample)} - g\ CO_{2(blank)}}{g\ soil\ (oven\text{-}dry\ basis)} \times 0.2727 \times 100$$

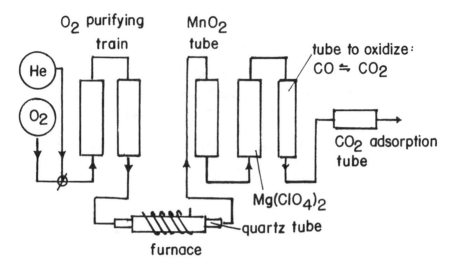

Figure 12.1 Schematic diagram of a dry combustion unit for the determination of total C_{org}.

12.2.2 Alternative Method

A variety of automated instruments are currently available for the determination of total carbon content in soils. One of these is the Perkin-Elmer 240C Carbon Analyzer (Perkin-Elmer Corp., Norwalk, CN). This analyzes not only total soil carbon, but also H and N at the same time, using the same principles as are used in the dry combustion method. The Carbon Analyzer is fully automatic.

Determination of Soil Organic Matter

Reagents
1. Helium, He, gas
2. Oxygen, O_2, gas
3. Copperoxide, CuO (granulated)
4. Copper, Cu (granulated)
5. Silver vanadate and silver wool
6. Magnesium perchlorate, $Mg(ClO_4)_2$
7. Soda-lime

Procedure

Weigh 500 mg of sample in a Pt boat and place this sample in a quartz tube. The analysis is started by cleaning the combustion unit from CO_2 gas by sweeping it with a He flow for 2 min. Calcareous soils should be treated first with H_2SO_4, as discussed previously. The sample is then heated at 1000°C for 5 min under a constant flow of oxygen. The mixture of gases produced, CO, CO_2, N_2, NO and H_2O, are passed over CuO to convert the CO into CO_2. The S and halogen gases are removed by passing the gas through silver vanadate on Ag wool. The gases then flow into a tube containing Cu granules sandwiched between Ag wool. This reduction tube is heated to 650°C. The NO gas is reduced into N_2 by the Cu granules, which are converted into CuO. The gas volume is adjusted to constant pressure and temperature in a mixing chamber. Then the gases are allowed to flow through a sequence of detectors and traps for the removal of H_2O and CO_2. The first detector measures the total gas composition. Water is then removed by a magnesium perchlorate trap before the gas flow passes detector 2, which measures the H content. After CO_2 is removed in a soda-lime trap, the gas flow passes the last detector, which measures the C content. The remaining gas in the mixture is N_2.

12.3 DETERMINATION OF OXIDIZABLE ORGANIC C

Many methods are available for the determination of oxidizable organic carbon by a wet combustion procedure. It is called wet combustion because oxidation occurs in an aqueous solution with the assistance of an oxidant, such as $KMnO_4$, or $K_2Cr_2O_7$. The chromic acid method, introduced by Schollenberger (1927, 1945), is the most common method today. The analysis uses solid $K_2Cr_2O_7$ or a $K_2Cr_2O_7$ solution in the presence of H_2SO_4, or a mixture of H_2SO_4 and H_3PO_4, with or without external heating. The use of $K_2Cr_2O_7$ with a mixture of H_2SO_4 and H_3PO_4 (Allison, 1960, 1965) is reported to yield a complete oxidation, by which all carbonaceous compounds are converted into CO_2:

$$2\ Cr_2O_7^{2-} + 16\ H^+ + 3\ C \rightarrow 4\ Cr^{3+} + 8\ H_2O + 3\ CO_2$$

The amount of CO_2 produced is measured, and the value is used as a measurement of the total C_{org} content. However, frequently the "fresh" organic residue remains unaffected, and the result in terms of total organic carbon refers only to C from decomposed organic matter, charcoal, and part of the fresh organic residue.

The other chromic acid oxidation method, using $K_2Cr_2O_7$ and H_2SO_4, measures only the decomposed fraction of soil organic matter or the humus. This method, known also as the *Walkley-Black method* (Walkley, 1947; Peech et al., 1947), uses a known excess of $K_2Cr_2O_7$. Part of the $K_2Cr_2O_7$ is consumed in the oxidation process and converted into Cr^{3+} ions as can be noted in the reaction above. The unused part is determined by *back* titration. By difference from a blank titration, the used part of $K_2Cr_2O_7$ is known and applied in the calculation of C_{org} content. Since two methods for determination of total C_{org} have been presented above, the following sections will discuss the determination of C_{org} from the humus fraction by the Walkley-Black method.

Determination of Soil Organic Matter

12.3.1 Walkley-Black Wet Combustion Method

Reagents
1. Potassium dichromate, $K_2Cr_2O_7$, $1\ N$ (= $1/6\ M$): Weigh 49.00 g $K_2Cr_2O_7$ in a 1 L volumetric flask. Dissolve with 800 mL tilled water, and make up to volume.
2. Sulfuric acid, H_2SO_4 96%, reagent grade.
3. Ferrous sulfate, $FeSO_4.7H_2O$, reagent grade, $0.5\ N$: Weigh 140 g $FeSO_4 \cdot 7H_2O$ in a 1 L volumetric flask. Dissolve in 800 mL distilled water under constant stirring. Add 15 mL of concentrated H_2SO_4, cool the mixture, and make up to volume with distilled water.
4. Ferroin indicator. This indicator is available as a solution from the Fisher Sci. Co., Fair Lawn, NJ, under the trade name, *1,10-Phenanthroline Ferrous Sulfate Complex*.

Procedure
Weigh 500.0 mg of soil, that has passed a 100 mesh sieve, in a 500-mL Erlenmeyer flask. Add with a burette 10 mL 1 N $K_2Cr_2O_7$ solution. Swirl to mix. Add carefully from a dispenser 20 mL concentrated H_2SO_4. Extreme caution should be exercised in dispensing the acid, since H_2SO_4 is very dangerous. Swirl gently to mix, while holding the mouth of the Erlenmeyer flask away from your body. Do not touch the bottom of the flask, since the reaction creates enormous amounts of heat. Allow the solution to cool to room temperature (20-30 minutes), after which 20 mL distilled water are added. Swirl again to mix. Add 5 drops of ferroin indicator, and titrate the excess (unused part) of chromic acid with ferrous sulfate or ferrous ammonium sulfate to a clear blue-brown endpoint (T mL). Run a blank using the same procedure (B mL).

Calculations

% C_{org} = (B - T) x N x 3 x 1.14 x (100/mg oven dry soil)

% Organic matter = % C_{org} x 100/58

12.4 DETERMINATION OF HUMIC MATTER

A variety of methods may be used to extract humic substances from soils. Ideally the extractant should meet the following requirements: (1) the reagent should not produce changes in the physical and chemical nature of the substances extracted, and (2) the reagent should be able to extract quantitatively all the humic substances from soils. Over the years, both inorganic and organic extractants have been evaluated for their effectiveness in isolating humic compounds (Stevenson, 1965[a]; Schnitzer et al, 1959; Tan, 1982), usually with mixed results in meeting both the requirements stated above. Some of the reagents, such as dilute bases, are capable of quantitatively removing all the humic fractions. However, they are believed to have some influence on modifying the physical and/or chemical properties of the extracted substances (Flaig et al, 1975), and the possibility of creating artifacts still confronts the investigator. Some of the reported inorganic extractants are listed in Table 12.1. Of the reagents listed in the table, NaOH and $Na_4P_2O_7$ are the most widely employed in the extraction of humic matter. Introduced for the first time in 1919 by Oden, NaOH appears to be the most effective in the quantitative removal of humic matter in soils. Currently, this reagent is recommended by the International Humic Substances Society for the extraction of humic acids (Calderoni and Schnitzer, 1984). However, it is believed that NaOH may induce autoxidation of humic acids during extraction. Therefore, many scientists suggest that extraction with NaOH be conducted

Determination of Humic Matter

Table 12.1. Inorganic Reagents Used in Extraction of Humic Matter

Acids	Bases and salts
0.1 M HCl	0.1 M NaOH
0.025 M HF	0.5 M NaOH
1% H_3BO_3	0.1 M Na_2CO_3
	0.1 M NaF
	0.1 M $Na_4P_2O_7$, pH 7.0
	0.1 M $Na_4P_2O_7$, pH 9 to 10
	0.2 M Na_2-EDTA
	0.1 M $Na_2B_4O_7$

under a N_2 gas atmosphere (Schnitzer, 1982). On the other hand, Bremner (1950) reported that the use of a N_2 atmosphere had no significant effect on the NaOH extraction of humic acids. This finding is supported by Tan et al. (1991), who found no differences in chemical and spectral properties of humic acids extracted by the NaOH method under air or under a N_2 atmosphere.

Generally, a milder solution of NaOH (0.1 M instead of 0.5M) is preferred for extraction purposes. It is believed that for qualitative analyses, a very weak NaOH solution (0.01 to 0.001 M) is even better than 0.1 M NaOH.

Sodium pyrophosphate ($Na_4P_2O_7$), although not as effective as NaOH, is used frequently for the extraction of humic matter from soils high in sesquioxide contents (Kononova, 1961). To increase its effectiveness in extraction, a solution with a pH of 9 to 10 is recommended. Although reports to the contrary are present, the use of 0.1 M pyrophosphate often eliminates the need to decalcify calcareous soil samples prior to extraction, as is required with the use of NaOH. The disadvantage of using $Na_4P_2O_7$ is that the extracted material is difficult to purify.

Extraction with acids, as proposed by Schnitzer et al. (1959), technically yields only fulvic acids.

The organic solvents used for extraction of humic matter are oxalic acid, formic acid, phenol, benzene, chloroform, or mixtures of these, and acetylacetone, hexamethylenetetramine, dodecylsulfate, and urea (Schnitzer and Khan, 1972). Thus far none of these has been satisfactory. Using 0.5 M and 0.1 M of hydroxymethylamine, Orioli and Curvetto (1980) obtained humic acids with different carboxyl contents than those extracted with NaOH.

In view of the above, the NaOH extraction procedure will be presented in the following sections.

12.4.1 Extraction of Soil Humic Matter

Reagents

1. Sodium hydroxide, NaOH, 0.1 M: Dissolve 4.0 g NaOH in 1 L of distilled water.
2. Hydrochloric acid, HCl, 4 M.
3. Hydrochloric acid and hydrofluoric acid mixture, HCl + HF: Mix 5 mL of concentrated HCl with 5 mL of 52% HF in a 1 L polyethylene volumetric flask. Dilute with distilled water to the mark.
4. Amberlite XAD-8 resin (40-60 mesh, Rohm and Haas).
5. Dowex 50-X8, H-saturated, cation exchange resin. Packed the dowex resin as a water slurry in a 50-cm long chromatographic tube. Leach it several times with 0.1 N HCl, and wash it with distilled water until Cl⁻ free.

Determination of Humic Matter

Procedure

Weigh 10 g of soil (< 2.0 mm) in a propylene centrifuge tube. Add 50 mL 0.1 M NaOH, and shake the mixture for 24 h with a wrist-action shaker (if calcareous soils are used, wash them with HCl before the extraction process with NaOH). After shaking the mixture, collect the dark colored supernatant solution by centrifugation at 10,000 rpm for 15 min. Wash the soil residue once with 50 mL distilled water and add the colored liquid, collected by centrifugation, to the supernatant. Then centrifuge the combined colored humic solution at 15,000 rpm for 15 min to ensure complete removal of fine colloidal clays. Discard the precipitate, whereas the humic solution is then acidified to pH 2.0 by adding HCl in order to precipitate the humic acid (HA) fraction. Separate the supernatant containing fulvic acid (FA) and the precipitated HA by centrifugation at 15,000 rpm for 15 min. Both the HA and FA fractions are collected for further purification as discussed below.

Purification of the HA Fraction

Dissolve the HA precipitate obtained above with 0.1 M NaOH, and centrifuge it at 10,000 rpm for 15 min. Discard the undissolved fraction. Reacidify the dissolved fraction to a pH of 2.0 with HCl, centrifuge at 10,000 rpm for 10 min, and discard the liquid supernatant. Shake the HA precipitate at the bottom of the tube loose with 50 mL HCl + HF mixture in order to reduce the ash and Si content. Centrifuge at 10,000 rpm for 5 min, and discard the HCl+HF extract, but save the HA precipitate. Next, wash the HA precipitate thoroughly with distilled water, centrifuge, and discard the wash-water. Redissolve the HA precipitate with 50 mL 0.1 N NaOH. This HA solution is then purified further by allowing it to flow through a H-saturated cation exchange (Dowex 50-X8) column, after which it is freeze-dried, weighed and stored in an amber-colored flask. The HA flowing from the cation exchange column has a pH between 2.0 and 3.0, because it is highly protonated.

Notwithstanding this low pH, the HA still remains in solution, though unstable solution (Lakatos et al., 1977). Such a behavior of HA corrresponds to that of aquatic HA, which stays in solution in the Okefenokee swamp water with a pH of 3.8 (Tan, 1993).

Purification of the FA Fraction

The colored supernatant containing the FA fraction is purified by passing it through an amberlite XAD-8 column. See the next section (12.4.2) on *extraction of aquatic humic matter* for the preparation of the XAD column. Fulvic acid, retained by the XAD-column, is washed twice with distilled water to remove carbohydrates and other extraneous compounds. The adsorbed FA fraction is then eluted from the column with 0.1 M NaOH, and allowed to flow through a H-saturated cation exchange (Dowex 50-X8) column for final purification. The purified FA is freeze-dried, weighed and stored in an amber-colored flask.

12.4.2 Extraction of Aquatic Humic Matter

Today it is known that HA and FA exist not only in soils, but also in streams, lakes, swamps and oceans. Many of the Coastal Plain streams and swamps of the southeastern United States and other parts of the world carry dark colored water. The dark color is generally attributed to humic matter in solution, which is composed mostly of FA (Tan et al., 1990). The isolation of aquatic humic matter from black water requires a somewhat different procedure than the extraction of HA from soils (Tan et al., 1989).

Reagents

1. Amberlite XAD-8 resin (40-60 mesh; Rohm and Haas). The resin is cleaned by shaking it successively with 0.1 M NaOH and methanol. Before the resin is packed in a polyethylene chromatographic column, it is rinsed several times with

Determination of Humic Matter

distilled water. After packing, the column is leached alternately with 0.1 M NaOH and 0.1 M HCl. This leaching is repeated twice (Thurman and Malcolm, 1981).
2. Sodium hydroxide, NaOH, 0.1 M (see extraction of soil humic matter, section 12.4.1).
3. Sodium hydroxide, NaOH, 0.2 M : Dissolve 8 g NaOH in 1 L of distilled water.
4. Hydrochloric acid, HCl (see extraction of soil humic matter, section 12.4.1)..
5. HCl + HF mixture (see extraction of soil humic matter, section 12.4.1).
6. H-saturated cation exchange (Dowex 50-X8): See extraction of soil humic matter (section 12.4.1).

Procedure

Five to 10 L of black water, depending on the humic matter concentration, are filtered through a micropore filter (0.45 µm) to remove suspended inorganic matter. The filtered black water is then acidified with HCl to pH 2.0, and allowed to flow through the XAD-8 resin column at a flow rate of 200 mL/h. The humic substances retained by the column are washed thoroughly with distilled water, and eluted with 0.2 M NaOH. The dark colored solution is collected, and acidified with HCl to pH 1.0. This process will precipitate the HA fraction, while the FA fraction remains in solution. The HA precipitate is separated from the FA solution by centrifugation at 10,000 rpm for 15 m, purified by twice dissolution and re-precipitation, and washed once with a HCl + HF mixture. After dialysis against distilled water for 4 nights, the purified HA is freeze-dried, weighed and stored.

In order to concentrate the FA fraction, the FA solution is passed again through the XAD-8 column. The adsorbed FA is washed with distilled water, and eluted with 0.1 M NaOH. It is then allowed to flow through a H-saturated cation exchange column (Dowex 50-X8) for a final purification. The purified FA is freeze-dried, weighed and stored.

12.4.3 Determination of Total Acidity of Humic Matter

The total acidity of humic matter is attributed to the presence of dissociable protons or H⁺ ions in aromatic and aliphatic carboxyl and phenolic hydroxyl groups. The value of total acidity is used as an index of the negative charge or the cation exchange capacity of the humic compound. Several methods are available for its determination, e.g., $Ba(OH)_2$, KOH, NaOH, and Ba acetate methods (Tan, 1982). Currently the $Ba(OH)_2$ method is used widely. In this procedure, the humic matter sample is saturated with a known excess of Ba^{2+}. The unused fraction of $Ba(OH)_2$ is back-titrated with a standard acid (Tan, 1982; Schnitzer, 1982).

__Barium hydroxide Method__

Reagents
1. Barium hydroxide, $Ba(OH)_2$, 0.2 N: Dissolve 34.2 g $Ba(OH)_2$ in 1 L of CO_2-free distilled water. Protect the $Ba(OH)_2$ solution from CO_2 gas with a soda-lime trap.
2. Hydrochloric acid, HCl, 0.2 N: Pipet 15.9 mL conc. HCl (37%) into a 1-L volumetric flask, and make up to volume with distilled water. Standardize the normality before use.

Procedure
Weigh 20.0 mg of finely ground humic matter in a 125-mL ground-stoppered Erlenmeyer flask. Add 10 mL 0.2 N $Ba(OH)_2$ solution. Displace the air in the flask by bubbling N_2 gas in the mixture, close airtight, and shake for 24 hours with a wrist-action shaker at room temperature. Filter the mixture into a clean Erlenmeyer flask, and wash the filtered residue once with CO_2-free distilled water using a jet-spray. Titrate the solution potentiometrically with a standardized (0.2 N) HCl to pH 8.4 (=

Determination of Total Acidity of HA

T mL). Use a microburet for dispensing the standard acid. Run a blank (= B mL).

Calculations

me total acidity/g HA = (B - T) x N x (1000/weight of HA)

Approximate Range of Total Acidity Values

The value for total acidity is dependent on the method of analysis. The $Ba(OH)_2$ method usually yields higher values than the other methods (Tan, 1982). The average total acidity of humic acids, determined by the $Ba(OH)_2$ method, is 5 to 7 me/g (Table 12.2). The humic acid of some soils, such as the spodosols, is high in total acidity. Fulvic acids have been reported to exhibit higher total acidity values than humic acids (Schnitzer and Khan, 1972). According to Schnitzer and Khan, the carboxyl content of fulvic acid is also two to three times higher than that of humic acid.

12.4.4 Determination of Carboxyl and Phenolic-OH Groups

As indicated previously, the sum of carboxyl and phenolic-OH contents equals the total acidity. The total acidity reflects the negative charge or the cation exchange capacity of the humic compound.

A number of methods, using various processes and reagents, are available for the determination of carboxyl groups in humic matter. Among these methods are those using ion exchange, decarboxylation, iodometry, esterification, or calcium acetate (Tan, 1982). Many scientists use the calcium acetate method, which is based on the formation of acetic acid. This reaction may be expressed as follows:

Table 12.2. Average Total Acidity, Carboxyl and Phenolic-OH Contents in Humic Matter (Tan, 1992)

Humic matter	Total acidity	COOH	Phenolic-OH
Humic Acid		me/g	
Terrestrial	7.0	4.3	2.7
Aquatic	7.0	4.4	2.6
Geologic	5.8	3.8	2.0
Fulvic Acid			
Terrestrial	6.8	4.7	2.1
Aquatic	7.0	4.4	2.6
Geologic	9.4	8.2	0.8

$$2R\text{-}COOH + Ca(CH_3COO)_2 \rightarrow (RCOO)_2Ca + 2CH_3COOH$$
humic matter acetic acid

The carboxyl content is then determined by titration of the acetic acid with a standard base.

Calcium Acetate Method

Reagents
1. Calcium acetate, $Ca(CH_3COO)_2$, 0.2 N: Weigh 31.6 g calcium acetate in a 1-L volumetric flask, add 800 mL distilled water to dissolve under constant stirring, and make up the volume

Determination of Total Acidity of HA

with distilled water.
2. Sodium hydroxide, NaOH, 0.1 N: Dissolve 4 g of NaOH in 1 L distilled water. This solution must be standardized before use.

Procedure

Weigh 20.0 mg of ground humic matter in a 125-mL ground-stoppered Erlenmeyer flask. Add 10 mL of the calcium acetate solution, and dilute with 40 mL CO_2-free distilled water. After shaking the flask for 24 h with a wrist-action shaker, filter the mixture into a clean Erlenmeyer flask. Wash the residue with CO_2-free distilled water, and add the wash-water to the filtered solution. Titrate the solution potentiometrically with the standard NaOH solution to a pH of 9.0 (= T mL). Use a microburet for dispensing the standard base. Run a blank analysis (= B mL).

Calculations

me COOH/g = (T - B) x N x (1000/mg oven-dry soil)

The phenolic-OH concentration is determined by the difference between the total acidity and the COOH content:

me phenolic-OH/g = me total acidity/g - me COOH/g

CHAPTER 13

SPECTROPHOTOMETRY AND COLORIMETRY

13.1 DEFINITION AND PRINCIPLES

Spectrophotometric and colorimetric analysis are defined as a group of analyses based upon the measurement of the amount of light absorbed or emitted by a substance. The difference between the two methods is only in the level of sophistication, though this has been eliminated somewhat with the development of current modern instrumentation. According to the traditional concept, colorimetry makes use of natural or visible light. With the assistance of suitable light filters, light of a definite color (narrow range of wavelength) can be produced by the colorimeter. On the other hand, in spectrophotometry, light of a definite wavelength is produced by a set of *monochromators*, hence measurements can also be made in the ultraviolet and infrared range, where ordinary colorimetric measurements are impossible.

The terms spectrophotometry and spectroscopy are frequently used interchangeably. Spectroscopy usually refers to analysis in which a flame or torch is applied, whereas in spectrophotometry the primary source of radiation is provided by a light bulb.

Two basic groups of spectrophotometry can be distinguished:

Spectrophotometry and Colorimetry

(1) *absorption spectrophotometry*, and (2) *emission spectrophotometry*. Absorption spectrophotometry is concerned with measurement of light absorbed by a substance, and will be discussed in more detail in this chapter. On the other hand, emission spectrophotometry measures the quality of the spectrum and the intensity of radiation emitted by a sample. Atoms or molecules from the sample, exited in a flame, arc, or spark produce emission spectra. The intensity of their radiation is proportional to the number of atoms, hence to the concentration of the substance containing those atoms. Emission spectroscopy will be the topic of Chapter 14.

13.1.1 Absorption Spectrophotometry

Absorption spectrophotometry is concerned with measurement of absorption of light in the ultraviolet (UV), visible, infrared (IR), and radio-wave region of the electromagnetic radiation spectrum. Therefore, depending upon the kind of radiation involved, the following types of spectrophotometry can be recognized: (1) UV spectrophotometry, (2) visible light spectrophotometry, (3) infrared spectrophotometry, and (4) NMR (nuclear magnetic resonance) spectrophotometry. Atomic absorption spectrophotometry also belongs to this category. However, since a flame or torch is applied in this type of analysis, the term atomic absorption spectroscopy can also be used.

Absorption spectrophotometry is distinguished further into (1) colorimetry or spectrophotometry, (2) turbidimetry, and (3) nephelometry. *Spectrophotometry* or *colorimetry* is based on the measurement of light absorbed by a colored solution, hence the name colorimetry. Generally, the intensity of the color is proportional to the concentration of the substance in solution. Therefore, by measurement of the intensity of the color with a spectrophotometer or colorimeter the concentration of the substance can be determined. If the substance to be determined is colorless, it can often be converted into a colored compound by the application of a suitable reaction. For example: P in solution is colorless, but in dilute solution P can be transformed into molybdophosphoric acid, which in reduced condition is blue in color (see determination of total P, section 10.2.2).

Turbidimetry involves measurement of the amount of light absorbed by a white suspension, whereas *nephelometry* is based on the determination of the amount of light scattered by a suspension. In both these methods, the substance under investigation must be transformed into an insoluble compound which remains in suspension. Measurement of the light absorbed or scattered by the suspension gives the concentration of the substance in suspension.

The main advantage of absorption spectrophotometry is that very small amounts of substances can be determined rapidly and accurately, when other procedures would yield relatively great errors with these minute amounts. This method, therefore, is especially suitable for measurements of micro and semimicro quantities of substances.

13.1.2 Principles of Light Absorption

Absorption spectrophotometry is generally controlled by basic principles or laws of light absorption. Based on these basic principles Lambert, Bouger, and Beer have formulated the now famous *law of Lambert and Bouger* and *law of Beer*. These two laws are frequently combined and are then called the *law of Lambert and Beer* or more accurately the *law of Lambert, Beer, and Bouger*. Spectrophotometry is based upon this law.

The following principles hold when a light beam falls on a homogeneous layer of a substance. Part of the radiation is reflected, part is absorbed, and part is transmitted. The substance can be in a solid, liquid, or gas form. The incoming light is generally called the incident light. The relationship between the intensity of incident light (I_0) and the intensities of reflected light (I_r), absorbed light (I_a), and transmitted light (I_t)

Spectrophotometry and Colorimetry

can be expressed as follows:

$$I_o = I_r + I_a + I_t$$

When comparing the intensities of the beams transmitted through the solution (containing the substance) and through the solvent, the effect of I_r is neglected in practice, and the relationship is changed into:

$$I_o = I_a + I_t$$

13.1.3 Law of Lambert and Bouger

This law, formulated by Bouger in 1729 and Lambert in 1760, indicates in essence that transmission and absorption of light is a function of thickness of the medium:

$$I_t = I_o \, a^{-1}$$

where a = factor related to the fraction of incident light transmitted by a layer of 1 cm thickness. This factor is often called the *transmission coefficient*.

Rearranging the equation above gives:

$$I_t/I_o = a^{-1} = T$$

T is called *transmittance*.

By rearranging again and taking the log, the equation above changes into:

$$I_o/I_t = 1/T$$

$$\log(I_o/I_t) = \log(1/T) = D$$

where D = *optical density* or *absorbance*. Sometimes the symbol A is assigned to absorbance.

13.1.4 Beer's Law

This law, formulated by Beer and Bernard in 1852, indicates that transmission and absorption of light by a medium is a function of concentration:

$$I_t = I_o\, a^{-c}$$

where a = transmission coefficient, and c = concentration.

13.1.5 Law of Lambert and Beer

The combination of the law of Lambert and Bouguer and the law of Beer gives the *law of Lambert and Beer*, the basis in absorption spectrophotometry:

$$I_t/I_o = a^{-lc}$$

or in log form:

$$\log (I_t/I_o) = \log a^{-lc}$$
$$= (-lc)\log a \quad \text{or}$$
$$\log (I_o/I_t) = (lc)\log a$$

Log a is also called the *extinction coefficient*, and is assigned the symbol ε. The term molar extinction or molar absorptivity is used when the concentration is expressed in moles/L. By using ε instead of log a, the formula above can be changed into the fundamental equation in spectrophotometry:

$$\log (I_o/I_t) = \varepsilon lc$$

which indicates that the optical density or absorbance is proportional to concentration (c), and thickness of sample (l). If a cuvette or sample holder of 1 cm thickness is used, then l = 1, and $\log (I_o/I_t) = \varepsilon c$, in other words, absorbance is directly proportional to concentration.

Conformity with Lambert and Beer's law gives a straight line if optical density or absorbance is plotted against concentration (Figure 13.1). However, if transmittance readings are plotted against concentration a curvilinear regression line is obtained (see Figure 13.1 B), because $T = I_t/I_o$, indicating that transmittance is not proportional to concentrations.

13.2 VISIBLE LIGHT AND UV SPECTROPHOTOMETERS

Instruments designed for measuring absorption of light by substances in the visible and ultraviolet region of the spectrum have been called photometers, spectrometers, absorptometers, colorimeters, and spectrophotometers. The essential components

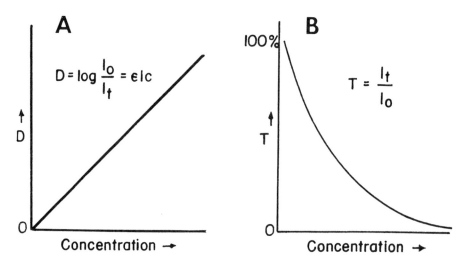

Figure 13.1 Relationship between: (A) optical density and concentration, and (B) transmittance and concentration.

of a spectrophotometer consist of a (1) stable source of radiation, (2) monochromator to resolve the radiation into component wavelengths or bands of different wavelength, (3) transparent sample holders, called tubes, cuvettes, or absorption cells, and (4) a radiation detector with associated readout system (Figure 13.2).

Spectrophotometry and Colorimetry

Single beam instrument

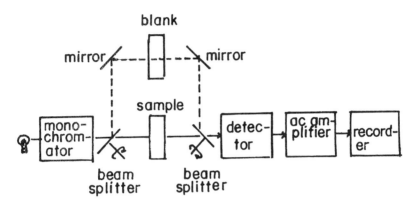

Double beam instrument

Figure 13.2 Schematic diagram of a component set-up in a single and double beam spectrophotometer.

13.2.1 Single Beam and Double Beam Spectrophotometers

Spectrophotometers can be distinguished into *single beam* and *double beam* instruments. Single beam instruments apply a single beam, as the name implies, in the analysis. The light beam passes through the cell containing the sample solution and on to the detector. In contrast, double beam instruments operate a beam splitter before the light beam enters the cell. One beam is directed through the blank or reference sample, and the other beam passes through the sample cell. The two beams are compared either continuously, or alternately many times a second. The latter causes the intensity of illumination in the photocell to change periodically. This produces an alternating current, the magnitude of which is determined by the amount of light transmitted by the sample solution and the blank. The concentration of the substance is then determined by (1) measuring the alternating current, or by (2) reducing the alternating current to zero by balancing the light beam with a photometer. The photometer reading determines the relative intensity, hence the concentration.

13.2.2 Source of Visible Radiation

A *tungsten filament lamp* is the traditional and least expensive source of visible and near infrared radiation. The filament is usually heated by a stabilized dc power supply, and emits radiation or light between 350 and 2500 mμ. A carbon arc provides a more intense visible radiation, but it is seldom used.

13.2.3 Source of Ultraviolet Radiation

The *hydrogen* and *deuterium lamps* are the most common sources of ultraviolet radiation in spectrophotometers. These lamps contain a pair of electrodes, enclosed in a glass tube with a quartz window. The tubes are filled with either hydrogen gas or deuterium gas at low pressure, hence the names hydrogen lamp or deuterium lamp. When a stabilized high voltage is applied to the electrodes, an electron discharge occurs which excites electrons in the gas molecules to high energy states. As these electrons return to their ground states, they emit radiation (light) in the region between 180 mμ and 350 mμ, the

Spectrophotometry and Colorimetry

ultraviolet region of the light spectrum.

A *xenon lamp* produces higher intensity UV radiation, but it also emits visible light, which appears as stray radiation in UV application. The UV radiation produced by the xenon lamp is also not as stable as that produced by the hydrogen or deuterium lamp.

13.2.4 Monochromators

Monochromators are light dispersing devices (Figure 13.3). The beam is split into several individual light beams with specific (narrow) wavelength. In scientific language, a monochromator is capable of resolving polychromatic radiation into its individual component wavelengths, and isolate them into beams with very narrow bands. The main component of a monochromator is either (1) a prism, or (2) a grating device. A prism disperses radiation into several types of light beams, from infrared (light with longer wavelength, λ) to UV (light with shorter λ). A quartz prism usually exhibits high dispersion. However, dispersion and resolution are poor near 700 mμ. In the infrared region prisms absorb light with longer wavelengths.

A *grating device* is a monochromator that consists of a highly reflective aluminized surface on which many parallel grooves (lines, trenches) are etched (Figure 13.4). Usually 600 to 2000 lines/mm are etched depending on the region of the spectrum for which they are intended. Grating monochromators are used mostly in UV and IR spectrophotometers. The incident light falling on each groove is diffracted over a range of angles. Reinforcement or destructive interference of the diffracted beams may occur. In *destructive interference* two diffracted waves cancel each other, whereas in *constructive reinforcement*, the two waves reinforce each other producing a wave exhibiting an amplitude that equals to the sum of the amplitudes of the two component waves. Constructive reinforcement obeys the *Bragg's law* of diffraction (see section 16.2.2).

MONOCHROMATOR

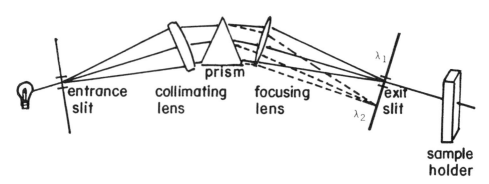

Figure 13.3 Schematic diagram of a monochromator

A perfect monochromator is considered to exhibit high resolution capacity and large dispersion. *Resolution* is the capacity to distinguish two adjacent (very close) absorption bands as separate spectral lines. *Dispersion* is a measure of linear spread between several spectral lines, in other words, the capability to split the incident beam. These split spectral lines, diffracted by a grating-type monochromator, are composed of first-order, second-order, and third-order diffractions (Figure 13.4), as stipulated by Bragg's law. Overlapping of these different orders frequently occurs. A first-order wavelength overlapping with several lower-order wavelengths may be eliminated by correct etching of the grating. The first-order diffraction can also be selected by rotating the grating device so that successive fractions of the first-order radiation falls on the exit slit.

Spectrophotometry and Colorimetry

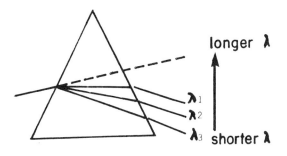

Prism type

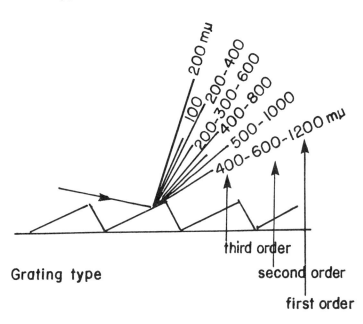

Grating type

Figure 13.4 Dispersion of polychromatic light into monochromatic light of different wavelengths (λ) by a prism type and a grating type monochromator

13.2.5 Detectors

A detector in spectrophotometry is a device that absorbs the energy of light and converts it into a measurable quantity of electrical current. The latter activates some type of recorder. The background signal created without light from the sample falling upon the detector is called the *detector noise*. These detectors are sometimes called *photoelectric detectors* and can be distinguished into (1) *phototubes*, (2) *photovoltaic cells*, and (3) *photomultiplier tubes*. For a detailed discussion of these tubes, reference is made to Willard et al. (1974).

Chapter 14

FLAME PHOTOMETRY

14.1 DEFINITION AND PRINCIPLES

In flame photometry a flame is used for (1) converting the sample from the liquid or solid state into the gas phase, (2) decomposing the sample into atoms, and/or (3) exciting these atoms into light emission. Therefore, three types of flame photometric methods can be distinguished: (1) flame emission spectroscopy, (2) atomic absorption spectroscopy, and (3) atomic fluorescence spectroscopy.

In all these methods, the sample is usually prepared in solution and sprayed into a flame, where it is desolvated, vaporized, or atomized. By using a solution spray a representative portion of the sample is distributed uniformly in the body of the flame. In *flame emission spectroscopy*, the atoms released by the sample are heated to an excitation level so that they emit radiation at the wavelength characteristic for that element. The intensity of the emitted radiation is considered proportional to the concentration of the atoms. However, since the intensity of emitted light is dependent on temperature, the sensitivity and reproducibility of the analysis vary also with temperature. In *atomic absorption spectroscopy*, the heat of the flame is controlled so that the atoms of the sample remain in an unexcited state. These unexcited atoms do not emit radiation on their own, but absorb radiation produced by a lamp from outside the flame. The intensity of absorbed radiation is a measure of the concentration of the atoms. *Atomic fluorescence spectroscopy* also involves irradiation of the atoms of the sample in the flame. Part of the incident light is absorbed and later emitted as *fluorescence* at characteristic wavelengths. The intensity of this fluorescence light is proportional to the concentration of the atoms.

Of the three, the most widely used methods are atomic absorption and flame emission spectroscopy. These two methods will be discussed below in more details. Although a flame is usually used, in many of today's instruments non-flame devices are becoming more common, e.g., graphite furnace in atomic absorption spectroscopy, and plasma atomization in plasma emission spectroscopy. With the graphite furnace, or also called the *heated graphite atomizer* (HGA), the sample is introduced as a liquid or solid and atomized in the graphite tube. In plasma atomization systems a plasma is used, which is in essence argon gas heated at high temperature behaving as a flame. The use of plasma emission is believed to have eliminated many of the problems associated with past emission sources.

14.2 ATOMIC ABSORPTION SPECTROSCOPY

Atomic absorption spectroscopy may be defined as a method for determining the concentration of an element in a sample by measuring the intensity of external radiation absorbed by atoms produced from a sample at a wavelength characteristic for that element. The basic principles of atomic absorption were discovered as long ago as 1860 by Kirchhoff, but it was not until 1955 that the analytical potential of atomic absorption was realized by Walsh, Alkimade, and Milatz.

14.2.1 Principles of Atomic Absorption

Atomic absorption refers to absorption of light at a specific wavelength by atoms or by an atomic vapor. The amount of light absorbed at the characteristic wavelength will increase as the number of atoms of the selected elements in the light path increase. The relationship between the amount of light absorbed and the concentration of the element present in known standards can be used to determine unknown concentrations in soil and plant extracts. In most atomic absorption spectrophotometers, instrument readout can be calibrated to display concentrations directly.

In principle this type of light absorption obeys the law of *Lambert and Beer*, which can be formulated as follows:

$$I_t = I_o \, e^{-alc}$$

where I_t = transmitted light, I_o = incident light, a = absorption coefficient of atomic vapor, l = path length of light beam, and c = concentration of atoms.

By rearranging and taking the log, the equation changes into:

$$\log I_t/I_o = \log e^{-alc} = -alc$$

or

$$\log I_o/I_t = alc = \text{absorbance}$$

The latter equation indicates that absorbance is a function of concentration and path length. As indicated in Chapter 13, by plotting absorbance against concentration a linear regression is obtained.

14.2.2 Instrumentation

The instrument consists of a primary source of radiation, a device to produce atomic vapor, a wavelength selector, a radiation detector and readout (Figure 14.1).

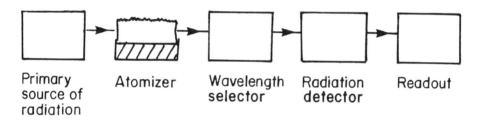

Figure 14.1 Schematic diagram of the main components of an atomic absorption spectrophotometer.

The relative positions of the main components are not necessarily to be as shown in the figure above. For example, the primary light source may pass through the wavelength selector before passing through the sample vapor (atomizer).

Primary Light Source

The primary sources of radiation in atomic absorption spectrophotometers are hollow cathode lamps. A hollow cathode

Flame Photometry

lamp is composed of a cathode and an anode sealed in a tube filled with an inert gas, argon, or neon (Figure 14.2). When a high voltage is applied across the electrodes, atoms of the filler gas become ionized and are attracted by the cathode. These ions hit the surface of the cathode with great velocity and dislodge atoms of the elements used to make the cathode. The displaced

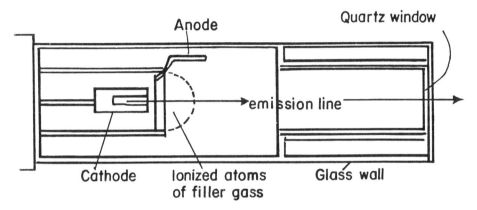

Figure 14.2 Hollow cathode lamp

atoms are excited and emit radiation at the characteristic wavelength of the element. For example, a Cu cathode emits a Cu spectrum, and a Zn cathode emits a Zn spectrum. Copper spectral lines can only be adsorbed by Cu atoms, whereas Zn spectral lines can only be adsorbed by Zn atoms. For this reason, a different hollow cathode tube must be used for each element to be analyzed. This type of a hollow cathode tube is called a *single element* tube. To avoid changing lamps in the analysis of several elements, several single element tubes are bundled together and rotated automatically after each reading (Figure 14.3). Another method is the application of a combin-

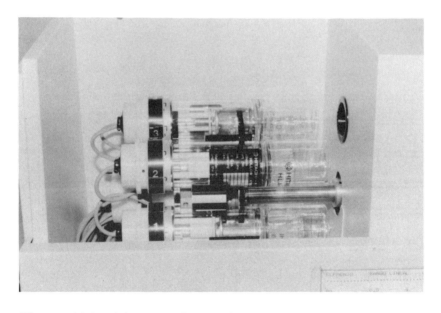

Figure 14.3 A battery of several single element hollow cathode tubes.

ation tube. By using two or more elements to form the cathode, a hollow cathode tube is produced that analyzes two or more elements. This type of lamp is called a *multi-element hollow cathode tube*, or a *combination tube*.

Polarized Light Source

Atomic absorption spectroscopy with the hollow cathode tube frequently yields interferences by background absorption. To correct for such kind of background interference, the primary beam is placed in a magnetic field of approximately 1 tesla, and the atomic spectrum is not only split into several components but also polarized. The light component polarized parallel to the magnetic flux is absorbed by the atom, whereas the light component polarized perpendicular to the magnetic flux is not. This effect, known as the *Zeeman effect*, was discovered in 1891 by Pieter Zeeman, a Dutch scientist. An example of a Zeeman polarizer covering the flame is shown in Figure 14.4. The primary single light beam is split by the Zeeman polarizer into the polarized *parallel* and *perpendicular* beam, respectively, providing in this way a double-beam spectrophotometric system. The parallel beam corresponds to the sample beam, whereas the perpendicular beam conforms to the reference beam. By using this method background correction and accuracy of analysis are enhanced considerably for many elements, as can be noticed from the analysis of Ca (Figure 14.5). However, it should be pointed out that the effect works with most elements, but with some elements, such as B and Cd, the difference in results by conventional versus Zeeman polarizing method is very small.

Atomizer and Flame

In most atomic absorption instruments the sample is atomized by a flame in successive stages. The sample entering the burning system is first nebulized, whereas the large drops are precipitated in the spray chamber and discarded through a flow system (Figure 14.6). The sample spray is then mixed, desolvated, and when it finally enters the flame decomposed into atoms. Constant feed rate of sample into the burner is critical. Flame shape is important. The flame should be narrow in width and should allow the incident light beam to meet all the

Figure 14.4 A Zeeman polarizer attachment in the flame compartment of an AA spectrophotometer.

the atoms under analysis so that the measurement reflects the concentration. The effective path length of the incident beam can be increased by multiple passages through the flame with a mirror system.

For an accurate analysis, the flame temperature needs to be

Flame Photometry

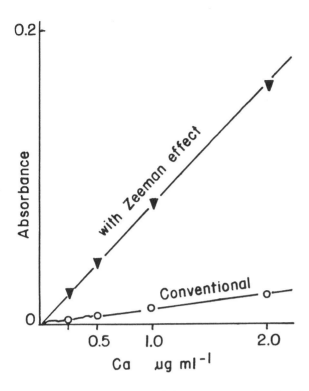

Figure 14.5 Analysis of Ca concentration by the conventional method and by using a Zeeman polarizer.

controlled properly. The temperature must not be too high to induce excitation of the atoms. The decomposition of the sample should yield neutral atoms capable of absorbing light. Some elements need a hot flame, while other elements need a cooler flame to break down into neutral atoms. The coolest flame in

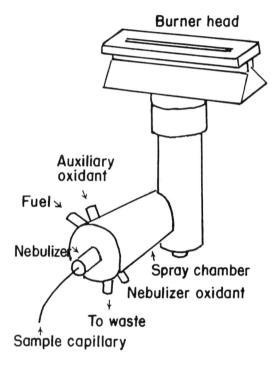

Figure 14.6 Schematic drawing of the burner system and atomizer.

atomic absorption is the air-propane flame that can reach a maximum temperature of 1800°C. It is sometimes used for elements that are very easy to atomize, such as Na, K, Cu and Zn. However the standard flame or most common flame used in atomic absorption is the air-acetylene flame. It is also considered a cool flame and can reach a maximum temperature of 2300°C. The hottest flame applied in atomic absorption is the nitrous oxide-acetylene flame which can reach a maximum temperature of 3000°C. This hot flame is required for elements that are difficult to atomize, such as Si, and Al. For example, the decomposition of $Ca(NO_3)_2$ in an air-propane flame is incomplete, as can be noticed from the following reactions:

$$Ca(NO_3)_2 \cdot H_2O \rightarrow Ca(NO_3)_2 + H_2O$$

$$2Ca(NO_3)_2 \rightarrow 2CaO + 4NO + 3O_2$$

The cool air-propane flame is unable to decompose the CaO further into its component Ca atom. In a hotter flame, the air-acetylene flame, the decomposition reaction proceeds one step further:

$$2CaO \rightleftharpoons 2Ca + O_2$$

However, the reaction above is an equilibrium reaction, hence the decomposition in atoms is incomplete. It will be complete in the hotter nitrous oxide-acetylene flame, resulting in the release of much more Ca atoms. Because of this, stronger analytical signals can be obtained making the analysis more accurate and reliable.

14.2.3 Samples and Sample Preparation

A large variety of samples in the solid, liquid and gas phase can be analyzed by atomic absorption spectroscopy.

1. **_Solid samples_** - Solid samples must first be dissolved, and the solution is analyzed. Soil, plant material, ores, and animal

tissue should be made soluble by appropriate fusion, extraction, and decomposition procedures as discussed in previous chapters. The solution or extract is then sprayed into the flame and analyzed.

The reagents used to dissolve solid materials should be selected with care. Generally, HNO_3 is suitable for flame and non-flame atomization, whereas H_2SO_4 and H_3PO_4 create sometimes interferences with flame atomization. Perchloric acid and HCl should be used with caution, since with Pb and Sn, it may form volatile chlorides that will be lost before atomization.

2. **_Liquid samples_** Liquid samples, such as blood, urine, cell-sap, wines, pollutants in water, etc., can be used directly as a spray and analyzed. However, frequently it is necessary to first digest liquid organic samples to remove the organic matrix compounds. Appropriate digestion procedures have been discussed in preceding chapters.

3. **_Gas samples_** Materials, such as metals, in the gas must be collected first by adsorption in a suitable reagent. This solution is then sprayed in the atomizer and analyzed.

14.2.4 Interferences

Atomic absorption and atomic emission methods are affected by changes in the concentration of ionized atoms. However, atomic absorption is usually not affected by radiation interference, but is influenced more by the solvent and by the chemistry of the system.

1. **_Chemical Interference_**

This type of interference is perhaps the most common interference in analysis by atomic absorption spectroscopy. As indicated earlier, in atomic absorption spectroscopy, the sample is sprayed in the flame with the purpose to decompose the substance into free atoms. However, some substances are difficult to decompose into neutral atoms. A mixture of neutral atoms and partly decomposed compounds are then produced in the flame, which cause interference, called *chemical interference*. Formation and decomposition of these difficultly dissociable compounds depend on many factors, such as flame temperature, composition of flame gas, ratio of fuel to oxidant gas, flame area for passage of light from the hollow cathode tube, concentration of element and accompanying elements, and aspiration efficiency of sample solution.

1.1 *Interference by Accompanying Elements*

Accompanying ions may cause interference by forming compounds that are very difficult to decompose in the flame. Frequently it is caused by the anion accompanying the cation to be determined. The anion affects the stability of the metal compound during the atomization process. For example, $CaCl_2$, when atomized, releases its Ca atom more easily than Ca-phosphate. The latter is more difficult to atomize. Phosphate ions are also known to decrease the absorption signal of Ca in an air-acetylene flame. Therefore, depending upon the accompanying anion, two solutions of equal Ca concentrations will yield different amounts of Ca atoms upon atomization. This, of course, results in differences in absorption. To eliminate such an interference, it is necessary to prepare calibration solutions from compounds with similar anions as the sample. The use of a NO_2 flame can also reduce such an interference. As discussed earlier, a higher temperature flame is capable of dissociating non-decomposable compounds into neutral atoms.

1.2 *Solvent Interference*

Materials in aqueous solution often give different results than those dissolved in organic solvents when analyzed by atomic absorption spectroscopy. It is believed that it is more difficult to atomize the metal or cation from an aqueous than from an organic solution. Though many factors are involved, it has been noticed that organic solvents increase the sensitivity of atomic absorption spectroscopic analysis. If the sample is dissolved in an organic solvent with lower surface tension than water, aspiration of the sample solution produces a finer spray. If the solvent burns easily, and its vapor pressure is higher than that of water, the increased rate of evaporation of the solvent may produce an increase in sensitivity in atomic absorption readings. However, benzene and toluene do not produce these effects. They will rather produce soot during combustion.

1.3 *Interference by Acids*

Phosphoric acid and sulfuric acid are believed to interfere with analysis of Ca and other alkaline earth metals in a low temperature flame. In the presence of phosphoric acid, Ca forms Ca-phosphate, which is very difficult to dissociate into neutral Ca atoms in a low temperature flame. In such a case, the use of a high temperature flame is recommended.

1.4 *Ionization Interference*

Analysis by atomic absorption spectroscopy is based on absorption of light by neutral (nonexcited) atoms. As discussed earlier, a temperature is required to dissociate the compound only into free neutral atoms. If the flame temperature is too high, some of the atoms, especially alkali and alkaline earth metals, tend to become ionized. The latter reduces the number of neutral atoms, and the accuracy of atomic absorption readings decreases considerably. To suppress this kind of interference, the best course is to add a high concentration of another element whose ionization potential is lower (larger ionization constant) than that of the atom under analysis.

Flame Photometry

2. *Physical Interference*

The rate of aspiration may change according to the properties of the solutes or solvents. This causes the sensitivity of analysis to change. Sample solutions with high concentrations of acids or coexisting elements exhibit high viscosity or surface tension, which can result in a poor spray efficiency. The latter will, of course, decrease the sensitivity of analysis. Compounds, containing high amounts of protein, such as blood or urine, usually display high viscosity.

Fluctuations in flame temperature may have a pronounced influence on the viscosity and surface tension of the sample solution. Elevating the solution temperature increases the aspiration volume.

14.3 ATOMIC EMISSION SPECTROSCOPY

Atomic emission spectroscopy may be defined as a method for determining the concentration of an element in a sample by measuring the intensity of radiation emitted by excited atoms produced from a sample at a wavelength characteristic for that element. No primary radiation source is needed with this method. The sample is decomposed into excited atoms that are capable of emitting radiation. The intensity of the emitted radiation is proportional to the concentration of the element.

14.3.1 Principles of Atomic Emission

Atomic emission refers to light emitted by atoms excited by the heat of a flame. The emission occurs only when sufficient thermal or electrical energy is applied to excite a free atom to an unstable state. Light is emitted when the atom or ion returns to the ground state (stable configuration). The wavelength of emitted light is characteristic of the element.

The emission spectrum is distinguished into a: (1) continuous spectrum, (2) band spectrum, and (3) line spectrum. A continuous spectrum is emitted by bright solids and consists of groups of lines. A band spectrum exhibits rather sharper defined lines than a continuous spectrum, whereas a line spectrum consists of even more sharply defined or spaced lines. Line spectra are produced by excited atoms that emit their extra energy as light at a definite wavelength.

The intensity of the emission spectrum at a given wavelength is proportional to the number of atoms. This number of excited atoms depends exponentially on the temperature, as formulated by the *Boltzmann equation*:

$$C_x = C_o \text{ exponential } (-E_j/kT)$$

where C_x = concentration of atoms at distance x, C_o = concentration of atoms in the bulk flame, E_j = excitation energy of excited state, k = Boltzmann constant (1.3085×10^{-16} erg/deg), and T = absolute temperature (°Kelvin).

If T increases, the negative exponent becomes smaller, hence C_o becomes larger. For example, if at 1000°K the number of atoms capable of emission equals 10^{-12}, then at 2000°K this number will become 10^{-6}. Hence, if the temperature increases twofold, the intensity of emission will increase one million fold ($\times 10^6$).

In order to produce a complete emission spectrum, the energy equal to the ionization potential must be absorbed by the atom. The *ionization potential* is defined as the negative log of the equilibrium constant, as explained below. Consider a compound AB to yield excited atoms A^+ and B^- for which the equilibrium reaction can be written as follows:

$$AB \rightleftharpoons A^+ + B^-$$

Flame Photometry

Application of the mass action law yield the following equilibrium constant, K,:

$$K = [(A^+)(B^-)]/(AB)$$

The compound AB is a pure solid, and hence at standard state its concentration equals unity. Therefore, at standard state the equation above assumes the form:

$$K = (A^+)(B^-)$$

By taking the -log, this equation changes into

$$-\log K = -\log (A^+)(B^-)$$

Since -log = p,

$$-\log K = pK = \text{ionization potential}$$

14.3.2 Emission Spectrographic Methods

The basic principle of light emission by metals burned in a flame, discovered long ago by Bunsen and Kirchhoff (Willard et al., 1974), was the basis for the development of the spectrograph. This instrument was for years the only tool for measurement of the intensity of emitted light in qualitative and quantitative elemental analysis. Currently, a variety of modern automatic instruments are available for this purpose. These different types of emission spectrographs are distinguished on the basis of the methods for excitation of atoms. The earliest energy source for excitation of atoms were ordinary flames. But with the development of modern technology, electrothermal sources such as dc arc, ac arc, and ac spark systems are employed currently. With the discovery of the laser beam technique, laser beam excitation of atoms is the latest addition today to the emission spectrographic methods. These methods are discussed in more detail below.

1. **_Flame Excitation_**, is a method in which a flame is used. It finds application in *flame photometry*, and in *inductively coupled plasma spectroscopy* (ICP). In ICP systems, a plasma is used, which is in fact heated argon gas which looks and behaves as a flame.

2. **_Excitation by Electrical Discharge_**, which is applied in *emission spectrographs*. Three types of electrical discharge commonly in use are (1) dc arc, (2) ac arc, and (3) ac spark. The dc arc method employs a voltage of 50 to 300 V, and introduces the sample into the electrical discharge. The heat generated by the discharge creates temperatures of 4000° to 8000°K, and vaporizes the sample into excited atoms. The ac arc system works according a similar principle as the dc arc method, but employs a considerably higher voltage (>1000 V) for the discharge. The ac spark is produced between two electrodes by an even higher voltage (10-50 kV) than that applied in the ac arc. Hence, it will result in much higher excitation energies than those produced by the dc or ac arcs. The ac spark is of advantage where high precision analyses are required. The spark method uses the solution as the conducting material in one of the electrodes. This solution, contained in a porous cup serving as an electrode, is made conductive by adding HCl. It is allowed to flow over the counter electrode when the discharge takes place. In some instruments the solution is placed in a

container below the counter electrode. A rotating metal or carbon disk is placed in the solution serving as the other electrode (Figure 14.7).

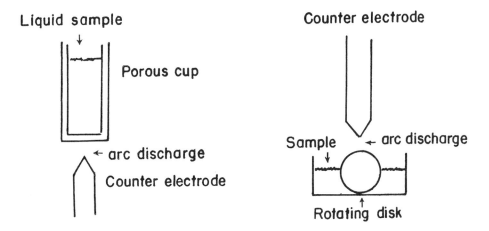

Figure 14.7 Electrical discharge by the ac spark method

3. *__Excitation by Radiation__*, such as applied in x-ray fluorescence spectrometry. Re-emission of radiation is called *fluorescence*.

4. *__Laser Microprobe Excitation__*. The laser microprobe method is applied in emission spectroscopy, and requires no sample preparation. A laser beam, directed to the sample, is absorbed. The intense heat vaporizes the sample into excited atoms capable of emitting radiation at their characteristic wave-

lengths. A sample spot of 50 µm can be selected.

It is beyond the scope of this book to discuss all the methods available in emission spectroscopy. Moreover, this chapter is entitled *flame photometry*, because it is the intention to discuss only the most frequently used method, the plasma emission spectroscopy, though the author realizes that the plasma is not exactly a flame. For the other types of emission spectroscopy reference is made to Willard et al. (1974).

14.3.3 Plasma Emission Spectroscopy

Plasma emission spectroscopy, also called *inductively coupled plasma* (ICP) spectroscopy, combines the principles of atomic absorption with spark emission spectroscopy.

The ICP is an argon gas plasma formed by a tangential flow of argon gas moving between the outer two quartz tubes of the ICP torch assembly (Figure 14.8). Electrical energy is applied through the heating coil and an oscillating magnetic field is formed. The plasma is created when the argon gas becomes conductive after exposure to the discharge of the coil creating charged particles (electrons and argon ions). Inside the induced magnetic field, the charged particles are forced to flow in a closed annular path. As they meet resistance to their flow, heating take place and additional ionization occurs. The plasma is then formed, which behaves and looks like a flame. This argon plasma can reach temperatures between 5500° to 8000°K, which are sufficient for a complete atomization of elements and minimizing chemical interference.

Sample preparation for analysis with ICP is similar as sample preparation for atomic absorption spectrophotometric analysis as discussed above. The sample is injected as an aerosol through the center of the plasma. Argon gas is used as the carrier gas for the sample. The difference with atomic absorption analysis is that in plasma emission spectroscopy the

Flame Photometry

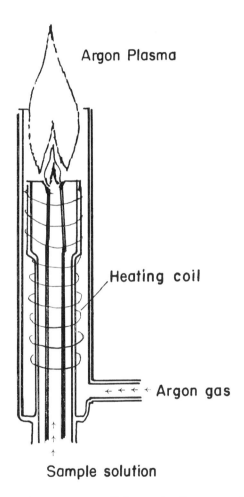

Figure 14.8 Schematic diagram of an ICP torch.

atoms released in the plasma are heated at a high state of excitation and emit radiation at their characteristic wavelengths. The intensity of emitted light is measured and used as an index of their concentration. Therefore, no primary radiation source, in the form of a hollow cathode tube, is needed.

14.4 Detection Limits of Atomic Absorption and Plasma Emission Spectroscopy

Most elements, except C, H, O, N, S, and P, can be analyzed with accuracy with either atomic absorption or plasma emission spectroscopy. Detection limits for most elements are perhaps slightly better for ICP emission than for flame atomic absorption spectroscopy. However, the differences are very small as can be noticed from Table 14.1. The detection limits listed in the table are within the 95% confidence level. The data in the table also indicate that the detection limits of atomic absorption using a graphite furnace far exceed those of ICP. However, some of the elements, such as B, U, W, and Zr, cannot be determined with the graphite furnace, and either flame AA or ICP should be used.

Flame Photometry

Table 14.1 Detection Limits for Atomic Absorption and ICP Spectroscopy (in mg/L)

Element	Flame AA	HGA	ICP
Aluminum Al	0.03	0.00001	0.02
Arsenic As	0.03	0.00015	
Barium Ba	0.14	0.0002	0.05
Boron B	0.70	0.015	0.004
Cadmium Cd	0.0005	0.000003	0.004
Calcium Ca	0.001	0.00005	
Chromium Cr	0.002	0.00001	0.005
Cobalt Co	0.006	0.00002	0.006
Copper Cu	0.001	0.00002	0.003
Gold Au	0.006	0.0001	
Iron Fe	0.003	0.00002	0.003
Lead Pb	0.010	0.00005	
Lithium Li	0.0005	0.0003	
Magnesium Mg	0.0001	0.000004	
Manganese Mn	0.0010	0.00001	0.001
Mercury Hg	0.17	0.002	
Molybdenum Mo	0.03	0.00002	0.008
Nickel Ni	0.004	0.002	0.01
Phosphorus P	53	0.03	0.05
Platinum Pt	0.04	0.0002	
Selenium Se	0.07	0.0005	0.05
Silicon Si	0.06	0.0001	
Silver Ag	0.0009	0.00005	
Tin Sn	0.11	0.0002	0.03
Titanium Ti	0.05	0.0005	0.002
Tungsten W	1.2		0.04
Uranium U	30		0.05
Vanadium V	0.04	0.0002	0.005
Zinc Zn	0.0008	0.000001	0.002

CHAPTER 15

INFRARED SPECTROSCOPY

15.1 DEFINITION AND PRINCIPLES

A considerable amount of information on the use of infrared spectroscopy in soil analysis is available in the literature. It is an important method of analysis in chemistry, pharmacy, industry, medicine, and forensics (Miller and Wilkins, 1952; Whetsel, 1968; Swinehart and Gore, 1968). Infrared spectroscopy has been applied in the study of clay minerals (Farmer, 1974; 1979; Ferraro, 1982; Stewart, 1970; Van der Marel and Beutelspacher, 1976) and in the qualitative analysis of functional group composition of humic matter (Stevenson, 1982; Tan and McCreery, 1970). Although the method has also been used to identify clay minerals and humic matter, a lot of criticism and doubts have been expressed on the results of such analyses (Mortenson et al., 1965). White and Roth (1982) believed that lack of access to infrared instruments and/or sympathetic colleagues with firsthand experience in infrared techniques to soil components are largely responsible for this situation. On the other hand, the current author is of the opinion that indictment as to the usefulness of infrared spectroscopy more probably finds its cause, in part in the complexity of the infrared spectra of clay mixtures and humic

Infrared Spectroscopy

preparations, and in part in questions of sample purity. The poor quality of many of the infrared spectra published on soil clays and humic matter has encouraged this, and has not contributed to the popularity of the method.

Infrared spectroscopy, also called infrared spectrometry or infrared spectrophotometry, is a spectrophotometric method that analyzes the absorption of infrared radiation by the sample under investigation. The infrared region of electromagnetic radiation, or infrared light, is located between the visible and the radiowave regions. As Table 15.1 illustrates, this area can

Table 15.1 Types of Infrared Radiation

Type	Wavelength	Frequency
	µm	cm^{-1}
Very near IR	0.75 - 2.5	13000 - 4000
Fundamental IR	2.50 - 25.0	4000 - 400
Far infrared	25 - 300	400 - 30

be divided into the very near or near, the fundamental, and the far infrared regions. Although the different types of radiation are characterized by the wavelength (λ) in µm, in infrared analysis it is customary to use the frequency in cm^{-1}. In this respect, the frequency, called the wavenumber ($\bar{v}$) or number of waves per cm, is chosen, instead of the frequency which refers to cycles/second. The wavenumber is the reciprocal value of the wavelength and accordingly this relationship may be expressed as:

$$\bar{v}\ (cm^{-1}) = 10^4/\lambda$$

The fundamental infrared radiation is usually applied in the infrared analysis of soil organic and inorganic constituents. The characteristic spectra of these compounds are found mostly between the frequencies 4000 cm^{-1} and 600 cm^{-1}.

15.2 ORIGIN OF SPECTRA

The absorption of infrared radiation depends on the vibrations of the atoms within the substance. Two types of molecular vibrations are important in the absorption of infrared light: (1) stretching vibrations, and (2) bending vibrations. These vibrations produce periodic displacement of the atoms with respect to another, resulting in a change in interatomic distance (Figures 15.1 and 15.2). When the vibrations are accompanied by a change in dipole moment, and when the frequency of vibration coincides with that of the infrared light, the vibration will absorb infrared radiation. It is then called *infrared active*. In contrast, vibrations of symmetrical molecules, not accompanied by a change in dipole moment, will not absorb infrared radiation, and are considered *infrared inactive*. The restoring force acting on the stretching vibration is believed to be greater than that in bending vibrations. Therefore, stretching vibrations will occur at higher frequencies than do bending vibrations (Adler et al., 1950).

The vibrational behavior of the atoms can be illustrated with the movement of a pair of spheres connected by a spring. The stiffness of the spring corresponds to the bond strength, whereas the masses of the spheres correspond to the masses of the atoms. If such a system is put in motion, the vibrations obey the *law of simple harmonic motion,* or *Hooke's law.* This law shows the relationship between the vibrational frequency, bond strength, and atomic masses as follows:

Infrared Spectroscopy

Stretching vibrations

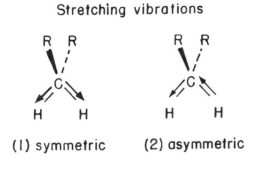

Bending vibrations

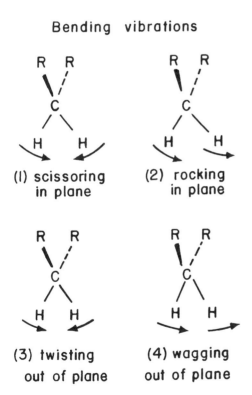

Figure 15.1 Major types of molecular stretching and bending vibrations between carbon and hydrogen atoms.

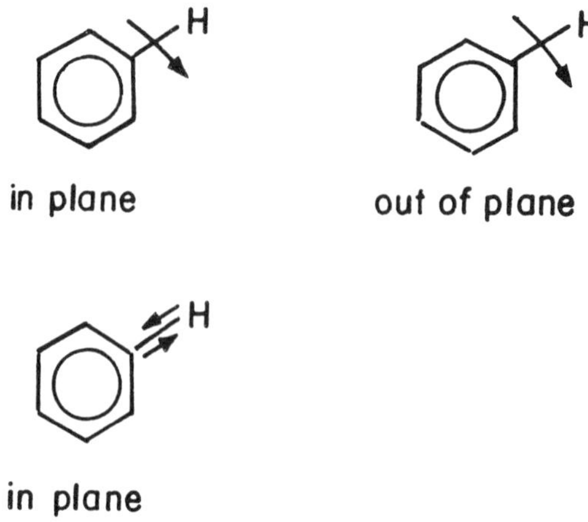

in plane   **out of plane**

in plane

Figure 15.2 Major types of vibrations of carbon and hydrogen in aromatic compounds.

$$\bar{v} = 1/2\ \pi c\sqrt{k(m_x + m_y)/(m_x + m_y)}$$

In this equation, $\bar{v}$ = frequency in cm^{-1}, c = velocity of light, k = force constant or bond strength, m_x and m_y = masses of atoms x and y, respectively. With this formula the frequency of vibration of a number of functional groups of inorganic and organic compounds can be calculated. The result can be used to verify the accuracy of an analysis or to identify the absorption bands of unknown compounds.

The infrared spectrum is composed of absorption bands, which is usually plotted with the wavenumber or frequencies in cm^{-1} as the abscissa against the percentages of transmittance as

Infrared Spectroscopy

the Y-axis. Although they look like *peaks,* the term *absorption bands* is common in infrared analysis. Some scientists prefer to use the percentage of absorption as the Y-axis, which yields an *upside-down* spectrum.

In spectrophotometry, transmittance, T, is defined as:

$$T = I_t/I_o$$

in which I_t = intensity of transmitted light and I_o = intensity of incident light.

According to the law of Lambert and Beer, T is related to the optical density or absorbance (D) as follows:

$$D = \log I_o/I_t = \varepsilon l c$$

In this equation, ε = extinction coefficient; l = optical path length, which corresponds to the thickness of sample holder; and c = concentration of sample.

Since $I_o/I_t = 1/T$, therefore:

$$D = \log 1/T = \varepsilon l c$$

This equation indicates that infrared spectroscopy can also be used for quantitative analysis. The latter is seldom applied in soil analysis, and merits further investigation. The main application so far of infrared analysis is for the qualitative determination of soil constituents.

15.3 INSTRUMENTATION AND SOURCES OF INFRARED RADIATION

Two groups of infrared spectrophotometers are generally recognized (Smith, 1979): (1) the sequential dispersive spectrophotometers, and (2) the multiplex nondispersive interferometer spectrophotometers. The latter group is also called the *Fourier transform infrared spectrophotometer*. The sequential dispersive spectrophotometers are mostly used in the qualitative determination of soil clays and soil humic matter. Double beam instruments that use gratings and filters as part of the monochromator are preferred.

15.3.1 Primary Source of Radiation

The primary source of radiation in these instruments is either the *globar* or the *Nernst glower*. The globar is an extremely stable source of infrared radiation. It is composed of a silicon carbide rod that is heated to approximately 1200°C by an electrical heating element. It emits continuous radiation between the wavelengths of 1 and 40 µm. The Nernst glower is a hollow rod of zirconium and yttrium oxides that is heated to approximately 1500°C. Although it is not as stable a source of infrared radiation as the globar, the Nernst glower is used in many infrared instruments.

15.3.2 Detectors

The detectors used to measure the intensity of radiant energy are frequently permanently evacuated thermocouples. In some instruments, bolometers, pneumatic or photodetectors are used. The detectors transform the energy received into an electrical current, which is amplified and plotted as curves. As indicated previously, the infrared spectrum is plotted with percentages of transmittance as functions of wavenumbers. In the most recent instruments, the addition of microprocessors has com-

Infrared Spectroscopy

puterized the infrared analysis, dramatically improving data output and handling.

15.4 SAMPLE PREPARATION

In general, samples for infrared analysis may be in gas, liquid, or solid form.

15.4.1 Gas Samples

Gas samples are analyzed in *infrared gas cells*. They consist of cylindrical glass tubes with NaCl, KBr or CaF_2 crystal windows. The path length varies from a centimeter to a few meters, by multiple passages and reflections.

15.4.2 Liquid Samples

Liquids are analyzed either as films of the pure compound (*neat*) or as thin solutions sandwiched between NaCl, KBr, or CaF_2 crystal plates. The plates are separated by a distance of 0.005 to 0.1 mm for neat liquids, and 0.1 to 1 mm for other solutions. Infrared cells must never come in contact with water, and should be cleaned only with organic solvent. Currently, infrared cells are available with *irtran windows*, which are insoluble in water. In older instruments, liquid samples are poured into rectangular cells provided with NaCl, KBr, or ThBr windows.

15.4.3 Solid Samples

Solids are examined by infrared spectroscopy as transparent pressed KBr discs. Sometimes solids are analyzed in a suspension form called *mull*. In this respect, the sample is made into a thick slurry by mixing it with a viscous liquid, e.g., *nujol* (parrafine oil), or chlorofluorcarbon grease. This mull technique is applicable only for qualitative analysis by infrared spectroscopy.

15.5 INFRARED ANALYSIS OF HUMIC MATTER

15.5.1 Procedures

Humic acid and fulvic acid are analyzed in the solid state as transparent KBr pellets. Weigh 1 mg of humic acid and 100 mg KBr (infrared grade). Because of its light color, 2 to 5 mg fulvic acid may be needed. Mix them by carefully grinding in an agate mortar. Transfer the mixture carefully into a die and press it into a pellet.

Two types of dies are available: (1) evacuable dies, which can be attached to a vacuum pump for removal of water vapor during pressing, and (2) non-evacuable dies, which provide for a rapid analysis. With the evacuable die, the die is evacuated to less than 2 cm Hg with a vacuum pump. One to 5 minutes of pumping is sufficient. With very dry powders 15 to 30 seconds may be sufficient. While the vacuum is maintained a pressure of 8 tons is applied to produce 13 mm pellets (1.25 tons for 5 mm pellets). For best result, use a hydraulic press. Release the pressure and vacuum after 1-2 minutes, and remove the KBr-HA pellet by pushing it out of the die with a plunger. The pellet is then transferred into a sample holder and placed in the infrared beam for analysis. It is helpful if all the KBr die components are kept warm (50°C). The pellet will then be warm when it is removed from the die. This will prevent the pellet from fogging even if the room temperature is high. If the pellet is below room temperature, it will fog immediately. All parts of the die must be kept clean when not in used, and they should be stored preferably in a desiccator.

When a non-evacuable die is used, the sample is squeezed between two polished screws, and the pellet is left inside the

Infrared Spectroscopy

die. The die serves as the sample holder and can be placed in the infrared instrument for infrared scanning.

15.5.2 Infrared Spectra of Humic Matter

The humic matter-KBr pellet is scanned from 4000 to 600 cm^{-1}. Sometimes scanning is continued to 400 cm^{-1}, but frequently the characteristic infrared bands are located within 4000 and 600 cm^{-1} (Table 15.2). The spectrum is often divided into two regions, a *group frequency region* (4000 -1300 cm^{-1}) and a *finger print region* (1300 - 650 cm^{-1}). In the group frequency region, the principal absorption bands may be assigned to vibration units that consist of only two atoms to a molecule. In the finger print region, single bond stretching and bending vibrations of polyatomic systems are major features. Molecules similar in structure may absorb similarly between 4000 - 1300 cm^{-1}, but will show differences in the fingerprint region.

Notwithstanding the many arguments on the usefulness of infrared spectroscopy in soil analysis, the method is capable of detecting and distinguishing between the different fractions of humic matter. The spectrum of humic acid is characterized by different infrared features than that of fulvic acid.

Fulvic Acid As can be noticed in Figure 15.3, the fulvic acid spectrum has a strong absorption band at 3400 cm^{-1}, a weak band between 2980 and 2920 cm^{-1}, a "shoulder" at 1720 cm^{-1} followed by a strong band at 1650 cm^{-1}, and a strong band at 1000 cm^{-1}. These bands are attributed to vibrations of OH, aliphatic C-H, carbonyl (C=O) followed by carboxyls in COO^- form, and ethyl, vinyl -CH-CH_2, aromatic aldehyde, amine and SH groups, respectively. This infrared spectrum shows close similarities with the infrared spectrum of soil polysaccharide.

Table 15.2 Infrared Absorption Bands of Functional Groups in Humic Matter (Tan, 1982;1994; Stevenson, 1982;1994)

Wavenumber cm^{-1}	Wavelength μm	Proposed assignment
3400-3300	3.94-3.03	O-H and N-H stretch
3380	2.950	Hydrogen bonded OH
2985	3.35	CH$_3$ and CH$_2$ stretch
2940-2900	3.40-3.44	Aliphatic C-H stretch
1725-1720	5.79-5.81	C=O stretch of COOH groups
1650-1630	6.00-6.10	C=O stretch (amide I), aromatic C=C, hydrogen bonded C=O, double bond conjugated with carbonyl and COO$^-$ vibrations
1650-1613	6.00-6.19	COO$^-$ symmetrical stretch
1460	6.85	Aliphatic C-H, CC-H$_3$
1440	6.95	C-H stretch of methyl groups
1435	6.97	C-H bending
1400	7.14	COO$^-$ antisymmetrical stretch
1390	7.20	Salts of COOH
1280-1230	7.80-8.10	C-O stretch, aromatic C-O, C-O ester linkage, and phenolic C-OH
1170-950	8.50-10.5	C-C, C-OH, C-O-C typical of glucosidic linkages, Si-O impurities, C-O stretch of polysaccharides
1035	9.67	O-CH$_3$ vibrations
840	11.9	Aromatic C-H vibrations

Infrared Spectroscopy

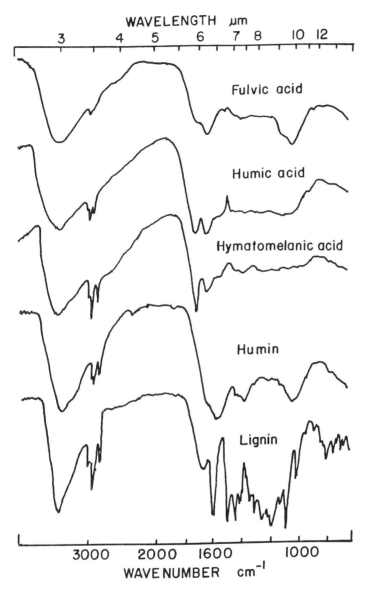

Figure 15.3 Characteristic infrared spectra of major humic compounds, and lignin.

Humic Acid In contrast to the above, the HA spectrum is characterized by a strong aliphatic C-H absorption band between 2980 and 2920 cm^{-1} and two strong absorption bands at 1720 and 1650 cm^{-1} for carbonyls and carboxyls in COO$^-$, respectively. In addition, the HA spectra lack the strong band at 1000 cm^{-1}. This feature frequently distinguishes it from the FA spectrum. The presence of a band at 1000 cm^{-1} in the HA spectrum is frequently associated with impurities in SiO_2. These impurities can be removed by washing the HA specimen with a dilute HCl-HF mixture. Some humic acids, especially those extracted from Ultisols, may not exhibit the two bands at 1720 and 1650 cm^{-1} respectively, but may have a spectrum with only one strong band for carboxyl groups in COO$^-$ form at 1650 cm^{-1}.

Hymatomelanic Acid The infrared spectrum of hymatomelanic acid, the ethanol soluble fraction of humic acid, has a very strong absorption band between 2980-2920 cm^{-1}, and at 1750 cm^{-1}, attributed to aliphatic C-H and C=O stretching vibrations, respectively. Tan and McCreery (1970) and Tan (1975) reported that hymatomelanic acid was an ester compound formed from humic acid and polysaccharide. They showed evidence through methylation of humic acids that the C-H group belonged to the polysaccharide, which was esterified to the carboxyl group of the humic acid molecule.

Humin Humin, the insoluble fraction of humic matter, exhibits an infrared spectrum, that closely resembles the infrared spectrum of FA. However, the humin spectrum is distinguished from the FA spectrum by a stronger aliphatic C-H absorption.

Lignin Lignin has also a spectrum that distinguishes it from HA or FA. Humic matter is believed to be a decomposition product of lignoid or lignin-like compounds. The only close similarity between the lignin spectrum and the spectrum of HA is noticed in the infrared features in the fingerprint region.

15.5.3 Methylation of Humic Acids

As indicated in the preceding section infrared analysis is very useful in the identification of the major humic compounds. The whole spectrum serves as a fingerprint and identifies the type of humic substance. Details about specific band assignments can also be obtained by studying spectra of humic matter derivatives produced by pretreatments, such as methylation (Tan, 1975; Tan and McCreery, 1970; Schnitzer, 1974), and acetylation (Stevenson, 1994) of humic and fulvic acids. These treatments are essentially adding methyl or acetyl groups respectively to the humic molecule, resulting in an increase in intensities of the bands at 2900 cm^{-1}, 1720 cm^{-1}, and at 1460 cm^{-1}, for C-H stretching, C=O stretching, and C-H deformation of CH_3 groups respectively (Table 15.3). Several methylation procedures are available in the literature, and one of the most

Table 15.3 Infrared Absorption Bands of Methylated Humic Acids (Tan, 1975; Tan and McCreery, 1970; Clark and Tan, 1969; Schnitzer and Skinner, 1965; Stevenson, 1994)

Wavenumber cm^{-1}	Band assignment
2920	Increased C-H stretching of added CH_3 groups
1720	Increased C=O (carbonyl) stretching
1650-1620	Decreased carboxyl in COO^{-1} stretching vibration
1440	Increased H-bending of added CH_3 groups
1220	Increased C-O vibrations of methyl esters

simple procedures, adapted from Schnitzer and Skinner (1965), is presented below.

Methylation Procedure Twenty mg of purified humic acid (or fulvic acid) are shaken for 48 h with 50 mL alcohol (90%), 50 mg silver oxide and 2 mL methyl iodide. The soluble fraction (methylated HA) is collected by centrifugation at 10 000 rpm for 15 min and freeze-dried before analysis by infrared spectroscopy.

15.6 INFRARED ANALYSIS OF SOIL CLAYS

15.6.1 Materials and Pretreatments

The total clay fraction (< 0.2 µm), coarse clay (2.0 - 0.2 µm) or fine clay (< 0.2 µm) can be used. Usually the total clay fraction is sufficient for a general identification of clay minerals. The clay samples must be free of organic matter, because organic matter may obscure the spectra of clays. Organic matter can be removed by treating the soil with H_2O_2 prior to particle size analysis, or by treating the clay fraction itself with this reagent. For specific purposes, the clay can also be deferrated by the dithionite-citrate-bicarbonate (DCB) method before infrared analysis. The chemical procedures of the pretreatments discussed above have been discussed in preceding chapters. Other pretreatments may be applied if required. Soil samples, especially B horizon samples low in organic matter, can be used directly.

15.6.2 Procedures

The clay samples can be prepared in a solid or in a dispersed phase. Several methods have been proposed for mounting these samples in infrared analysis, e.g., (1) mull method, (2) clay film technique, or (3) KBr pellet technique. The clay film technique, using demountable cells and/or other support material, is done for samples in the dispersed phase. Currently, the use of clay films has attracted increased attention. However, the most simple and widely used method is the KBr pellet technique. Both these techniques are presented below.

__KBr Pellet Technique__ One mg of clay is weighed and mixed with 100 mg of KBr by grinding carefully but thoroughly in an agate mortar. Proceed according to the procedure described above for humic acid. For good results, two pellets may have to be made for use in the analysis of one sample. One pellet containing 2 mg (or more) clay and 100 mg KBr must be prepared to yield a spectrum with a good resolution in the group frequency region (4000 - 1300 cm^{-1}). This pellet is usually too concentrated for infrared scanning in the fingerprint region (1300 - 650 cm^{-1}). Therefore, a second pellet containing 1 or 0.5 mg clay mixed with 100 mg KBr should be prepared.

__Clay Film Technique__ Clay samples can also be prepared in a suspension form. Weigh 10 mg of clay and add 2 mL water. Disperse it ultrasonically in a beaker. Then pipet a sample onto an irtran-II window (disc), so that 1 mg/cm^2 or 5 mg/cm^2 of clay are transferred onto the cells. Place the irtran window in a desiccator above $CaCl_2$ and dry the sample at room temperature for 2 nights before infrared scanning. For a good resolution, the thickness of the clay film should be found experimentally, since it may differ with the type of clay.

The author has noticed that on many occasions clay films prepared on infrared cells, such as irtran-II (ZnS crystal) or NaCl crystals, yield better infrared resolutions than clay samples prepared by the KBr pellet technique. This will be discussed in more details in the next section. However, the disadvantage of the clay film technique is that it takes more time and it is more difficult to prepare a clay film than to make a pellet.

15.6.3 Infrared Spectra of Soil Clays

The different types of clays can be distinguished by different types of infrared spectra. The differences are mainly observed in the regions between 4000 and 3000 cm^{-1}, and 1800 to 600 cm^{-1}. Between 3000 and 1800 cm^{-1}, the infrared spectrum of clays is usually featureless (Table 15.4).

Table 15.4 Infrared Absorption Bands of Functional Groups in Soil Clays (Tan, 1994)

Wavenumber cm^{-1}	Proposed assignment
3800 - 3600	OH vibrations from octahedrons in silicate minerals
3600 - 3250	O-Al-OH vibrations in sesquioxide clays
3400	OH vibrations from adsorbed water
1200 - 1150	O-Al-OH vibrations from octahedrons
1020 - 1000	Si-O vibrations, Si-O-Si antisymmetrical stretch
950 - 900	Al-OH vibrations

Kaolinite   As can be noticed in Figure 15.4, kaolinite is characterized by two strong bands between 3800 and 3600 cm^{-1}, when the sample is prepared by the KBr pellet technique. These bands are attributed to infrared absorption by octahedral OH stretching vibrations. An additional third and very sharp band is present at 3670 cm^{-1} when the sample is mounted as films on irtran windows. Using the irtran cells, the finger print or lower frequency region exhibits sharp bands for kaolinite at 1150 cm^{-1}, 1080 cm^{-1}, and at 910 to 920 cm^{-1} for O-Al-OH

Infrared Spectroscopy

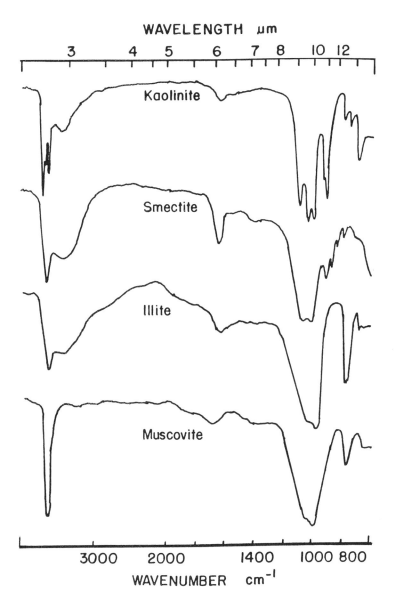

Figure 15.4 Characteristic infrared features of of kaolinite, smectite (montmorillonite), illite, and muscovite.

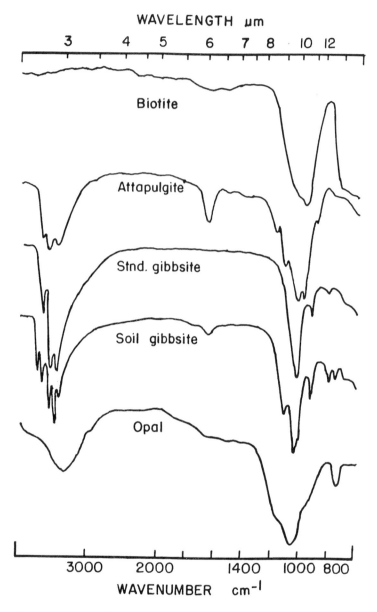

Figure 15.5 Characteristic infrared features of biotite, attapulgite, reference (stnd.) gibbsite, soil gibbsite, and opal.

vibrations. In addition, the spectrum shows a sharp band at 1020 cm^{-1} for Si-O vibrations. The bands at 1080 and 1020 cm^{-1} are segregated as a very weakly duplet in most analysis by the KBr pellet method reported for kaolinite.

Smectite The KBr curve of smectite is characterized by a single sharp band at 3640 cm^{-1}, followed by a broad band between 3600 to 3400 cm^{-1} for OH stretching and water, respectively (Figure 15.4). In the lower frequency region, smectite has also a strong band at 1050 cm^{-1} for Si-O vibrations. The strong band at 1600 cm^{-1} is the water band. However, using clay films on irtran-II, the band at 3640 cm^{-1} for OH-stretching vibrations and at 1050 cm^{-1} for Si-O vibrations become very strong and sharp. In addition, the bands at 1150, 910, 880, and 850 cm^{-1} also increase considerably in intensity in clay film samples.

Illite Illite has an infrared spectrum which is slightly different from the spectrum of smectite (Figure 15.4). The spectrum of illite has broader bands than that of smectite for OH stretching vibrations at 3600 cm^{-1}.

Muscovite and Biotite Muscovite and biotite, minerals from which illite is believed to have been formed, can be distinguished from illite by their different infrared spectra. The spectrum of muscovite exhibits a single sharp band for OH stretching vibration at 3610 cm^{-1} (Figure 15.4). This band is absent in the biotite spectrum (Figure 15.5).

Attapulgite Attapulgite has a spectrum (Figure 15.5), which is also different than the others, although its spectrum shares some similarities with the infrared features of kaolinite and smectite.

Gibbsite Gibbsite is characterized by an infrared spectrum, which is especially different from that of kaolinite or smectite in the region between 4000 and 3400 cm^{-1} (Figure 15.5). Standard gibbsite (purchased from the Wards Scientific Establishment Co.) exhibits a band at 3620 cm^{-1}, which is followed by two bands (duplet) at 3540 and at 3480 cm^{-1}. In the low frequency region, gibbsite shows only one dominant peak at 1030 cm^{-1} for O-Al-OH vibration. This peak is the reason for determining the kaolinite band at 1080 cm^{-1} as a separate independent band, rather than combining it with the 1020 cm^{-1} band and considering them as a duplet.

However, the infrared spectrum of soil gibbsite differs slightly from that of the standard gibbsite. Soil gibbsite has three bands (a triplet) between 3600 and 3400 cm^{-1} for O-Al-OH vibrations, instead of two bands as in standard gibbsite. This soil gibbsite was present in the clay fraction of a Hayesville Bt2, a clayey, kaolinitic, mesic Typic Kanhapludult. The infrared bands between 3800 and 3620 cm^{-1} in the Hayesville clay sample are caused by an admixture with kaolinite.

Opal Opal is a form of biogenic quartz with the formula $SiO_2 \cdot nH_2O$. Precious opal is composed of silica spheres in an ordered packing. The infrared spectrum (Figure 15.5) is characterized by a water band at 3400 cm^{-1}, and a strong band at approximately 1000 cm^{-1} for Si-O stretching vibrations, corresponding to the formula of opal stated above.

CHAPTER 16

X-RAY DIFFRACTION ANALYSIS

16.1 DEFINITION AND PRINCIPLES

In x-ray diffraction analysis, x-rays are used to determine crystalline compounds in soils. The analysis is based on the principle that x-rays will be diffracted by crystal planes. Soil minerals can be identified by measuring this diffraction at several different angles. The diffraction pattern serves as a *fingerprint* of each mineral.

X-ray radiation, a form of electromagnetic radiation, was discovered in 1895 by Wilhelm Konrad Roentgen (Pauling, 1964; Whittig and Allardice, 1986). Therefore, it is called *roentgen radiation* in Europe. Roentgen or x-ray radiation is located between the wavelengths of 200 Å to 0.02 Å (20 nm to 0.002 nm) in the electromagnetic spectrum. It is a form of highly energetic white light, and can be divided into soft x-ray, x-ray and hard x-ray. The analytical region of the x-ray spectrum is between the wavelengths of 20 Å to 0.2 Å.

X-rays are generated after ejection of an electron from the K-shell of an atom by bombardment with a highly energetic electron beam (Figure 16.1). Continuous x-ray radiation is emitted after the subsequent transition of orbital electrons from states of high to low energy levels. If the transition takes place with an electron from the L-shell, Kα radiation is emitted.

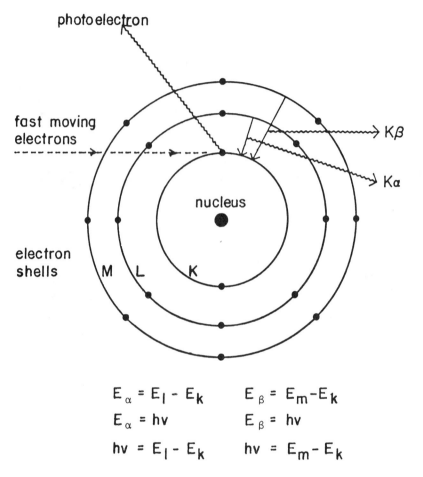

Figure 16.1 Principles in the production of Kα and Kß radiation, two principal forms of x-rays.

X-ray Diffraction Analysis

When an electron from the M-shell occupies the vacated position in the K-shell, Kß is emitted. Therefore, the emitted x-ray is composed of both Kα and Kß radiation. For x-ray diffraction analysis the Kα radiation is used, and the Kß is blocked by a filter system.

When a beam of x-rays hits a sample, absorption, scattering or fluorescence of the x-rays may occur. These three reactions form the basis of three major groups of x-ray analyses (Table 16.1).

Absorption method finds application in *x-ray radiography*, which was originally used in medical science. Today this type of x-ray analysis is an important method in the industry and aerospace technique, where it provides a means for determining deformities and discontinuities in bulk materials. Recently, x-ray absorption methods have been used in security screening, e.g., for high speed examination of luggage. In the industry, the content of cans can be studied without opening the can. In soil and plant analysis, the presence of different labeled elements is indicated by patches on a photographic paper of varying intensities of gray.

The scattering effect is the basis for *x-ray diffraction analysis*. The interaction of x-ray with an electron of an atom in the sample may result in an *incoherent* or *coherent scattering* of the x-ray. The scattering is said to be incoherent (*Compton scatter*), if energy is lost in the collision or interaction. The latter will especially take place if the electron is loosely bound. The scattering is considered coherent (*Rayleigh scatter*), if no energy is lost in the collision. The coherently scattered x-ray exhibits the same wavelength as the incident x-ray. X-ray diffraction is a special case of coherent scattering of x-rays. The method was developed after the discovery of x-ray diffraction in 1913 (Jenkins et al., 1981).

Fluorescence is the basis of *x-ray fluorescence spectrometry*. When the x-ray interacts with the sample, x-ray is absorbed. The absorbing atoms in the sample become excited and emit x-

Table 16.1 Analytical Applications of X-ray Radiation

Interaction	Analytical method	Applications
Absorption	X-ray radiography	Medicine, industry, security system, aero-space, animal science and plant physiology
Diffraction	X-ray diffraction	Identification of soil and clay minerals, crystallography
Emission	X-ray fluorescence: (a) wavelength dispersive spectrometry (b) energy dispersive spectrometry	Tissue analysis in plant and animal samples, qualitative and quantitative microanalysis in tissue samples

rays with a wavelength characteristic of the emitting atoms. This process is called x-ray fluorescence, and the produced x-ray is known as *secondary x-ray*. X-ray fluorescence spectrometry was also developed after diffraction of x-ray was discovered. The method is very useful in the analysis of plant and animal tissue, and in the microanalysis of element content in the tissue. This latter type of analysis can be divided into (1) a wavelength dispersive analysis by x-rays, and (2) an energy dispersive analysis by x-rays, also known as EDAX, EDXRA or EDS. A relationship exists between energy or wavelength and atomic number (Jenkins et al., 1981). This relationship may be expressed as:

$$E/1.24 = 1/\lambda = k(Z-s)^2$$

In this equation, E = energy of the secondary x-ray (in keV), λ = wavelength (in nm), k and s = constants, corresponding to the spectral series of the emission lines, and Z = atomic number.

Because of the relationship above, isolation of peaks by EDAX allows the identification of elements in an undisturbed plant and animal tissue. The element concentration can be determined by the characteristic peak intensity.

16.2 INSTRUMENTATION AND PRINCIPLES OF X-RAY DIFFRACTION

16.2.1 Sources of X-ray Radiation

An x-ray diffraction spectrometer usually consists of a generator, a water-cooled primary radiation source unit, a diffractometer, and a measuring electronic unit. X-ray diffraction instruments may vary slightly, depending upon the manufacturer. The major suppliers are Phillips Electronic Co., A.G. Siemens, General Electric Co., Hitachi, and Rigaku.

The primary source unit consists of a vacuum-sealed x-ray tube, contained in a water-cooled housing unit. The tube contains an anode, which is called the *target*, and a cathode (Figure 16.2). Numerous metals have been used as target

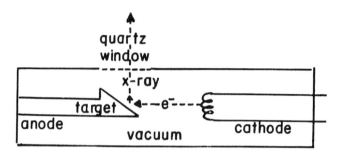

Figure 16.2 Schematic diagram of an x-ray tube (e⁻ = fast moving electrons hitting the target).

materials. The most common target elements are Cu, Fe, Mo and Cr. Most x-ray diffraction spectrometers are supplied with a Cu-tube. When a high voltage is applied across the cathode and anode, the electrons, released from the cathode, strike the surface of the target. Upon impact, energy is transferred to the metallic surface of the target, which then emits radiation in the 0.1 - 100 Å (0.01 - 10 nm) region. The wavelength of the x-ray emitted by the target depends upon the metal used and the applied voltage.

When all the energy (E) of the electrons is converted into radiant energy, the following condition holds:

$$E_{electrons} = E_{radiation}$$

$$E_{electrons} = eV \quad \text{and} \quad E_{radiation} = h\nu$$

X-ray Diffraction Analysis

in which e = valence, V = applied voltage, h = Max Planck's constant, and v = frequency.

$$v = c/\lambda$$

In this equation, c = the velocity of light in vacuum and λ = the wavelength.

Therefore: $hv = eV$

$$h(c/\lambda) = eV$$

$$\lambda = hc/eV$$

Since h, c and e are known, hce = 12,400. The final equation, characterizing the emitted x-rays, is then:

$$\lambda = 12{,}400/V$$

16.2.2 Principles of X-ray Diffraction

When such a beam of x-rays hits a crystal, x-rays are scattered among other things. A crystal plane is composed of a systematic and/or periodic arrangement of atoms in space. This kind of arrangement of atoms is called a *crystal lattice*. Each atom in the lattice serves as a scattering point. The coherently scattered waves may constructively interfere with each other, producing diffraction maxima. In order to meet the requirement for constructive interference, the scattered waves from each atom must be in phase. As illustrated in Figure 16.3, wave II

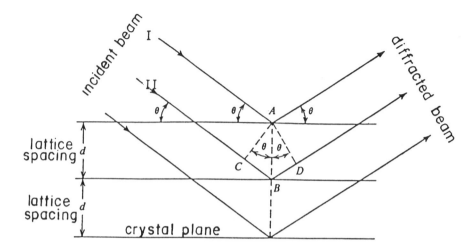

Figure 16.3 The conditions for diffractions of x-rays from crystal planes. The difference in path length between wave I and II is CBD=CB+BD. Since CB=BD, therefore CBD=2CB. Reinforcement occurs when 2CB=nλ (integral numbers of wavelengths). Since CB=ABsinθ, hence 2CB=2ABsinθ. Since AB=d, therefore, nλ=2dsinθ (*Bragg's law*).

must travel longer than wave I to leave the crystal plane. To ensure that the two waves remain in phase (or for reinforcement to occur), the path length difference between the two waves must equal a whole number (n) of wavelengths (λ). This condition for reinforcement was formulated by W.H. Bragg in 1913 (Bragg, 1933). Commonly known as *Bragg's law*, this

X-ray Diffraction Analysis

relationship may be expressed as:

$$n\lambda = 2d \sin\theta$$

in which n = integral numbers, λ = wavelength (Å), d = interplanar or lattice spacing, and θ = angle of incidence beam or angle of diffraction. Sometimes the lattice spacing is identified by the indices hkl. In this case, Bragg's law changes to:

$$n\lambda = 2d_{hkl} \sin\theta$$

The symbols *hkl* may refer to the 001, 010, 100 crystal planes, hence to the d_{001}, d_{010}, or d_{100} spacings.

Thus, x-ray diffraction will occur only if the Bragg's law is obeyed. The angle of the incidence beam (θ) is changed during analysis. When n = 1, the diffraction is called the *first-order diffraction*. By changing the angle of the incidence beam, n becomes 2 or 3, corresponding then to the *second-order* and *third-order* of diffractions, respectively. In clay mineral analysis, the x-ray spectrum, called *x-ray diffractogram*, is composed only of first- and second-order diffraction maxima.

Diffraction of x-rays from a succession of regularly spaced atoms produces a diffraction maximum, which is recorded. Because the interatomic distances in space are not the same between two minerals, the angles of diffraction will be very distinctive for a particular mineral. The interatomic distances in a crystal usually produce an array of diffraction maxima, which is used to identify the mineral.

16.3 QUALITATIVE X-RAY DIFFRACTION SPECTROMETRY

16.3.1 Pretreatments

For the identification of clay minerals, it is required to analyze the samples both in their original state and after they have been subjected to such pretreatments as deferration, saturation with K, saturation with Mg and glycolation, and heating at 500°C. Potassium saturation and glycolation of Mg-saturated samples are used to distinguish between expanding and non-expanding minerals (Table 16.2). Saturation with K will normally effect a collapse of intermicellar spacings, and the value for d-spacings of 20 - 17 Å exhibited by smectite may collapse to 10 Å. Reconstitution of the d-spacing to 17 Å is achieved by the glycolation procedure. None of these treatments will have any effect on the d-spacings of non-expanding minerals, e.g., kaolinite.

__Removal of Organic Matter__   Organic matter interferes in XRD, and should be removed to enhance the diffraction peaks. If organic matter has not been removed during the collection of clay in particle size analyzes, the clay samples should be treated with H_2O_2 according to the following procedure:

Weigh 1 g of clay, and add 5 mL distilled water followed by 10 mL H_2O_2 (30%). Digest the mixture on a hot plate at low heat. After the reaction has ceased completely, add 10 mL of water and centrifuge the mixture to remove the liquid. Wash the clay residue two times with 25 mL water and remove the liquid by centrifugation. The cleaned clay is dried, ground and stored for further analysis.

__Deferration__   In many cases, it is necessary to remove amorphous and free iron oxides from the clay samples prior to

Table 16.2 Effect of Pretreatments on the d-Spacings of Selected Clay Minerals

d-Spacings

Å	nm	Clay minerals
	K-saturated samples	
14	1.4	Vermiculite, chlorite
10-12	1.0-1.2	Smectite, illite
7.2-7.5	0.72-0.75	Halloysite, metahalloysite
7.15	0.715	Kaolinite, chlorite
	Mg-saturated samples	
14	1.4	Vermiculite, chlorite, smectite and illite
10-12	1.0-1.2	Illite, halloysite
7.2-7.5	0.72-0.75	kaolinite, chlorite
	Mg-saturation + glycolation	
17-18	1.7-1.8	Smectite
14	1.4	Vermiculite
10-12	1.0-1.2	Illite, halloysite
7.15	0.715	kaolinite
	Heating at 500°C (773 K)	
14	1.4	Chlorites
10	1.0	Vermiculites
7.0	0.7	Chlorites (kaolinite and sesquioxides become amorphous)

analysis by x-ray diffraction (XRD). Amorphous and free iron may be present as coatings and may interfere in the analysis by producing low-intensity diffraction peaks. The removal of these compounds, called deferration, may enhance the diffraction peaks. Deferration is usually performed by the *dithionite-citrate-bicarbonate (DCB) method* (Jackson, 1956). The procedure is as follows:

Weigh 1 or 5 g of clay in a centrifuge tube. Add 40 mL of 0.3 M Na-citrate and 5 mL 1 M NaHCO$_3$. Stir and place the tube in a waterbath. Heat the waterbath to 80°C. Add approximately 1 g of solid Na$_2$S$_2$O$_4$, and stir occasionally for 15 m at 80°C. At the end of the 15 m digestion period, add 10 mL of saturated NaCl solution to flocculate the clay. Centrifuge the mixture and collect both the supernatant and the deferrated clay.

The supernatant can be used for free and amorphous Fe, Al and Si content determination. The deferrated clay is used to make oriented glass slide preparations for XRD, which will be discussed below.

K-Saturation

Weigh 50 mg of clay in a polyethylene centrifuge tube. Add 10 mL of distilled water and a few drops of 0.5 M NaOH, and disperse the clay with a hand shaker. Add 10 mL 2 M KCl to flocculate the clay. Centrifuge and discard the supernatant. Wash the clay two times with 10 mL 1 M KCl to insure complete saturation with K. Centrifuge and discard the supernatant. Wash the excess KCl once with 50% methanol and once with 95% acetone. Centrifuge and discard the methanol and acetone. Disperse the clay again ultrasonically with distilled water. Prepare a glass slide sample for XRD analysis.

Mg-Saturation and Glycolation

Weigh 50 mg of clay in a polyethylene centrifuge tube. Add 10 mL of distilled water and a few drops 0.5 M NaOH, and disperse the clay with a handshaker. Add 0.2 N HCl dropwise under constant stirring to adjust the pH to 3.5-4.0. Add 10 mL 10 N Mg(OAc)$_2$ and shake with a handshaker. Centrifuge the suspension and discard the supernatant. Wash the clay two more times with Mg(OAc)$_2$, followed by a washing with 1 N MgCl$_2$ to insure complete saturation. Wash the excess salts out once with 50% methanol and once with 95% ethanol. Disperse the clay again ultrasonically with distilled water. Prepare a slide sample and dry the slide over CaCl$_2$ in a desiccator for analysis by XRD or for glycolation. Place the glass slide with the dry Mg-saturated clay upside down over ethylene glycol in a desiccator overnight. As soon as the dry color changes into a slightly moist color, the sample is ready for analysis by XRD.

Heating at 500°C After performing XRD of the K-saturated sample, the slide is heated in a muffle furnace at 500°C for at least 15 m. Cool, and store the slide over CaCl$_2$ in a desiccator. Kaolinite, halloysite, gibbsite, hematite, goethite and some HIV become amorphous to x-rays as a result of heating at 500°C. Frequently, heating is performed to distinguish between vermiculite, chlorite and illite. Vermiculite has interlayer hydroxy alumina complexes in its structure. These layers will be destroyed by heating at 500°C, with the consequent collapse and shift of the d-spacing of vermiculite from 14 Å to 10 Å. On the other hand, heating at 500°C will have no effect on chlorite, but will, however, produce a second-order 7.2 Å diffraction peak. The latter can be mistaken for kaolinite. However, heating at 500°C will make kaolinite amorphous to XRD, as indicated above, and both the 7.13 A and 3.56 A peaks of kaolinite will disappear.

16.3.2 Mounting Techniques of Samples in XRD

Many methods have been developed for the preparation of samples for analysis by XRD. Examples are the random powder method, random-orientated rod method, the wedge method, the glass slide method, and the ceramic plate method.

Random Powder Sample   In the random powder sample, the crystals lie in a random position to each other. In this method, the sample can be placed in a capillary tube, or with the aid of glycerol and gum tragacanth, the clay sample can also be made into a paste, and rolled into a rod of 0.3 to 0.5 mm thickness, which is analyzed by XRD. In the *wedge method*, the paste is pressed into specially designed wedge holders. These types of random powder samples are usually analyzed by a powder camera x-ray diffraction unit.

Random-Oriented Samples   Currently, the porous ceramic tile and the random-oriented glass slide method are preferred for mounting XRD samples. However, if sufficient amounts of samples are available, a random powder type of mounting using Al sample holders, instead of capillary tubes, may also be used, which is perhaps the most rapid and simple method. In the *ceramic tile method*, a tile is placed in a clay suspension, and the liquid is removed by suction so that the surface of the tile is covered by a thin layer of oriented clay. The clay can also be deposited on the tile surface by centrifugation. However, a more rapid and simple method is the oriented glass slide method. The specimens, mounted by these methods, are analyzed with a XRD unit that records the results as a graph, called previously a diffractogram.   The procedures for mounting samples by the oriented glass slide method and random powder method are presented below in more details.

16.3.3 Preparation of Oriented Glass Slide Samples

By this method a clay suspension has to be prepared, which is transferred on a microscopic slide producing random orientation upon drying, hence the name oriented glass slide sample.

X-ray Diffraction Analysis

Procedure Weigh 25 mg original or pretreated clay in a beaker. Add 2 to 5 mL H_2O and disperse the clay ultrasonically. Pipet the clay suspension onto one end of a clean microscopic glass slide, so that approximately 20 mg of clay is transferred per 10 cm^2. After the slide is dried over $CaCl_2$ in a desiccator, it is ready for analysis by XRD. The amount of clay transferred can be decreased or increased to produce a proper layer thickness. A thick layer of clay film tends to curl when dry.

The slide is placed in the holder (Figure 16.4) and usually scanned by XRD from 2° 2θ to 30° 2θ. Most of the characteristic first- and second-order diffraction peaks of clay minerals are located within this range.

16.3.4 Preparation of Random Powder Samples

As indicated above this method is perhaps the most convenient method in mounting samples for analysis by XRD. Ground clay can be used immediately in the dry and solid state, without dispersing it in water.

Procedure Weigh 100 mg of clay and fill the rectangular shallow hole in an Al specimen holder. Smooth the surface of the clay gently with a spatula so that the hole is covered with the sample. Clip the specimen holder in the space provided in the diffractometer and scan by XRD from 2° 2θ to 30° 2θ or higher.

16.3.5 Diffractograms of Clay Minerals

The diffractogram is composed of a series of diffraction peaks printed with the 2θ angles in the abscissa and radiation counts or peak intensities in the y-axis. The position of the diffraction peaks corresponds directly to the d-spacings.

The common unit for lattice spacing is the angstrom unit (1 Å = 10^{-10} m), which corresponds to the unit of x-ray wavelength.

Figure 16.4 Sample (slide) holder in a diffractometer.

However, the SI system requires the use of nanometers (1 Å = 0.1 nm). In older books, the kX unit is used, which is based on the effective spacing of cleavage planes of calcite (= 3.0290 kX). The kX units are easily converted into Å units by multiplying by 1.0020.

The value d(001)/n, called lattice spacing, can be measured from the results. If n = 1, the d(001)/1 represents the first-order diffraction spacings. If n = 2, then d(001)/2 represents the second-order diffraction spacing. The series of d/n values obtained, together with the intensities of the XRD peaks, serve

X-ray Diffraction Analysis

as fingerprints for the identification of the mineral species. Highest intensity of diffraction maxima is obtained from d(001) planes. The first-order d(001) diffraction peak, and in many cases together with the second-order diffraction peak, identify the mineral species. The following illustrations serve as a few examples.

Kaolinite   Kaolinite (Figure 16.5) exhibits a characteristic first-order diffraction at an angle of $2\theta = 12.4°$. The latter corresponds after conversion (Appendix D) to a d(001) spacing of 7.13 Å. The second-order diffraction is at $2\theta = 25.0°$, which corresponds to a d-spacing of 3.56 Å. These two peaks collapse after heating kaolinite at 500°C.

Smectite   Smectite or montmorillonite is characterized by a first-order x-ray diffraction peak of 12.3 Å which shifts to 17.3 Å after solvation or glycolation of the sample. Saturation with K produces a shift of the peak to 10 Å. The second-order diffraction peak is usually absent (Figure 16.6).

Illite   Illite exhibits a first-order diffraction peak of 10.1 Å. This peak will not collapse or shift after K- and Mg-saturation or glycolation of the sample (Figure 16.7). The intensity of this peak decreases somewhat upon heating the mineral at 500°C.

Vermiculite   Vermiculite has a characteristic first order peak at 14.3 Å (Figure 16.8), which shifts after K-saturation, and heating at 500°C to 10.1 Å. Mg-saturation+glycolation move the peak back to 14.3 Å.

Hydroxy-Al Interlayered Vermiculite (HIV)   Hydroxy-Al interlayered vermiculite is also characterized by a 14 Å peak (Figure 16.9). It is distinguished from vermiculite by the collapse of the 14 Å peak after heating HIV at 500°C. In addition, K- and Mg-saturation+glycolation produce no shift.

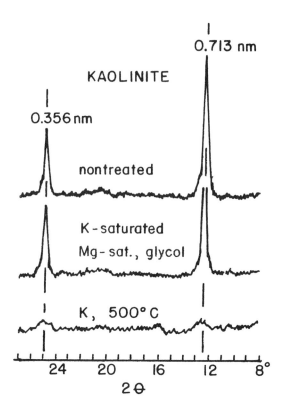

Figure 16.5 Identification of kaolinite using its characteristic XRD patterns (0.713 nm = 7.13 Å).

Gibbsite Gibbsite is identified by the dominant diffraction peak at 4.82 Å. Saturation with K and Mg, and/or glycolation has no effect, but the peak collapses by dissolution of the mineral with NaOH or heating at 500°C (Figure 16.10).

Goethite Goethite is easily recognized from the dominant peak at 4.12 Å (Figure 16.11). It reacts in a similar manner as gibbsite upon saturation with K and Mg, and/or glycolation,

X-ray Diffraction Analysis

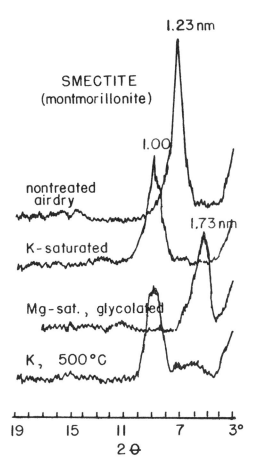

Figure 16.6 Identification of smectite (montmorillonite) using its characteristic XRD patterns (1.23 nm = 12.3 Å).

and heating at 500°C. However, dissolution of the mineral with NaOH has no effect on the diffraction peak, which is the differentiating characteristic with gibbsite.

For additional characteristic d-spacing values of other clay minerals, reference is made to Table 16.3. As can be noticed

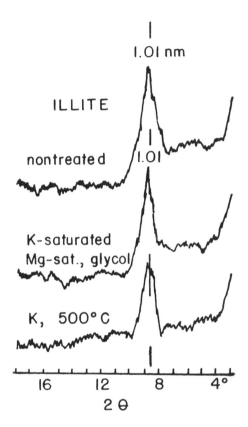

Figure 16.7 Identification of illite using its characteristic XRD patterns (1.01 nm = 10.1 Å).

from the d-spacings in Table 16.3, many of the minerals have similar or overlapping diffraction peaks. In such cases, additional analyses have to be performed using samples that have received the appropriate pretreatments (see Figures 16.5-16.11).

X-ray Diffraction Analysis

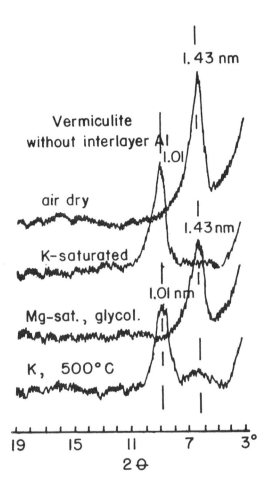

Figure 16.8 Identification of vermiculite using its characteristic XRD patterns (1.43 nm = 14.3 Å).

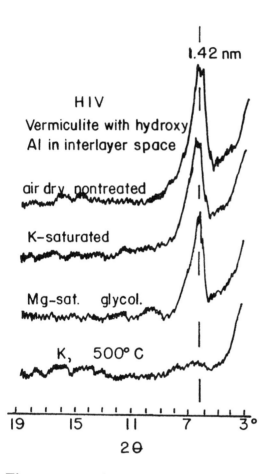

Figure 16.9 Identification of HIV (hydroxy-Al interlayered vermiculite) using its characteristic XRD patterns (1.42 nm = 14.2 Å).

X-ray Diffraction Analysis

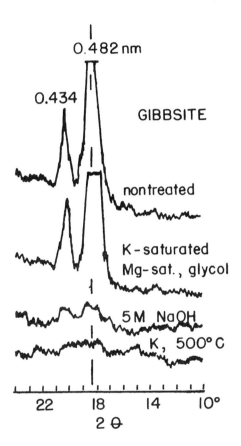

Figure 16.10 Identification of gibbsite using its characteristic XRD patterns (0.482 nm = 4.82 Å).

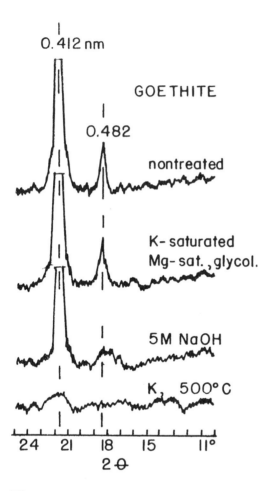

Figure 16.11 Identification of goethite using its characteristic XRD patterns (0.412 nm = 4.12 Å).

X-ray Diffraction Analysis

Table 16.3 Characteristic d-Spacing Values of Soil Minerals

(d/n)-Spacing Å	Intensity[a]	Probable mineral
32	-	Regularly interstratified chlorite and smectite
28	-	Regularly interstratified smectite and mica; chlorite + vermiculite
24	-	Regularly interstratified vermiculite and mica
17.7 - 17.0	(10)	Smectite, glycolated
14.0 - 15.0	(1)	Vermiculite, HIV
13.0 - 14.0	(5)	Chlorite
12.0 - 15.0	(9)	Smectite (airdry), vermiculite
11.4 - 11.7	(10)	Hydrobiotite
10.0 - 14.0		Hydrous mica
10.8	(10)	Halloysite, hydrated
10.2		Attapulgite
10.0	(10)	Mica, illite, smectite (K-saturated, 550°C)
9.20 - 9.40	(9)	Talc
9.10 - 9.20	(6)	Pyrophyllite
8.40 - 8.48		Amphibole
7.10 - 7.30	(6)	Antigorite
7.10 - 7.20	(6)	Chrysotile
7.20 - 7.50	(8)	Metahalloysite
7.10 - 7.20	(10)	Kaolinite, dickite, nacrite
7.00 - 7.20		Chlorite, vermiculite
6.30 - 6.45	(5)	Feldspar
6.25	(10)	Lepidocrocite
6.23	(10)	Boehmite
5.40	(7)	Mullite
5.42	(5)	Palygorskite

Table 16.3 (Continued)

(d/n)-Spacing Å	Intensity[a]	Probable mineral
5.0	(9)	Muscovite
4.96	(3)	Goethite
4.85	(3)	Magnetite
4.83	(10)	Gibbsite
4.72	(10)	Bayerite
4.70 - 4.80	(9)	Chlorite
4.62	(8)	Spinel
4.60 - 4.90	(5)	Vermiculite
4.60	(4)	Sepiolite
4.60 - 4.70	(6)	Talc
4.57		Pyrophillite
4.40 - 4.50	(9)	Illite, muscovite
4.49	(8)	Palygorskite
4.45 - 4.46	(4)	Kaolinite, dickite
4.42	(10)	Metahalloysite
4.40	(8)	Nacrite
4.35 - 4.36	(6)	Kaolinite, dickite
4.34		Gibbsite
4.29		Gypsum
4.26	(4)	Dickite
4.25 - 4.21	(7)	Quartz
4.20 - 4.30	(5)	Palygorskite
4.17	(6)	Kaolinite
4.15		Goethite
4.13	(6)	Dickite
4.12	(3)	Kaolinite
4.00 - 4.20	(8)	Feldspar
4.05	(10)	Crystoballite
3.98	(10)	Diaspore
3.84	(5)	Calcite, kaolinite

X-ray Diffraction Analysis

Table 16.3 (Continued)

(d/n)-Spacing Å	Intensity[a]	Probable mineral
3.80 - 3.90	(4)	Feldspar
3.78	(6)	Dickite
3.73 - 3.75	(6)	Feldspar
3.73	(7)	Ilmenite
3.72	(3)	Maghemite
3.69	(5)	Palygorskite
3.67	(7)	Hematite
3.59 - 3.60	(6)	Feldspar, chrysotyl
3.56 - 3.58	(10)	Kaolinite, dickite, nacrite, metahalloysite
3.57 - 3.58		Kaolinite, metahalloysite
3.54		Smectite, solvated
3.47		Anatase
3.44 - 3.48	(4)	Feldspar
3.40	(10)	Aragonite
3.39	(10)	Mullite
3.36	(3)	Pyrophyllite
3.36	(3)	Goethite
3.35	(10)	Quartz
3.33		Muscovite, sericite
3.30	(3)	Gibbsite
3.30	(10)	Calcite
3.28	(9)	Lepidocrocite
3.27	(5)	Aragonite
3.21 - 3.28		K-feldspar
3.20	(6)	Bayerite
3.16	(10)	Boehmite
3.15 - 3.06	(7)	Gypsum
3.15	(4)	Crystoballite
3.12 - 3.23		Plagioclase

Table 16.3 (Continued)

(d/n)-Spacing Å	Intensity[a]	Probable mineral
3.10 - 3.25		Hornblende
3.10 - 3.25	(8)	Feldspar
3.03	(10)	Calcite
2.88		Dolomite
2.87		Gypsum
2.83 - 2.77		Apatite
2.73		Hornblende
2.73 - 2.69		Apatite
2.69		Hematite
2.54		Ilmenite
2.53		Magnetite
2.52		Olivine
2.51		Hematite
2.46		Olivine
2.43		Goethite
2.34		Boehmite
2.28		Calcite
2.19		Dolomite
2.09	(2)	Calcite
1.92		Gypsum, calcite
1.88		Anatase

[a] Numbers in parentheses refer to intensity on a scale of 1 to 10 (with 1 representing the weakest intensity).

16.4 QUANTITATIVE X-RAY DIFFRACTION SPECTROMETRY

Although x-ray diffraction spectrometry is mostly conducted for the identification of crystalline minerals in soils, the method has also been used for the quantitative determination of mineral species (Whittig and Allardice, 1986). It is believed that the relative intensities of diffraction peaks provide a basis for measuring the concentrations of the minerals present. However, arguments have been presented that the diffraction intensity is affected not only by the concentration of the minerals present, but also by a number of other factors such as degree of orientation, particle size, chemical composition and crystal imperfection (Jackson, 1956). Of the several methods reported to compensate for the influence of these factors (Thomson et al., 1972; Hughes and Bohor, 1970; Brindley, 1980), the method using an internal standard has attracted the most attention. It involves the construction of a calibration curve from samples mixed with an internal standard. The ratio of the peak intensity of a standard soil mineral to the diffraction intensity of the internal standard is plotted against the weight ratio of the standard soil mineral to the internal standard. It is a rather time consuming and tedious procedure, since each component under investigation needs a standard curve. Recently a matrix-flushing method using corundum (α-Al_2O_3) as an internal standard has been proposed (Chung, 1974[a,b]). According to Chung, the method does not need a calibration curve, and no previous information about the approximate concentration range is required. This method will be presented below.

__Procedure__ Weigh 50 mg of the sample and mix with a known amount of corundum (α-Al_2O_3). A one-to-one weight ratio is suggested (Chung, 1974). In order to ensure optimum particle size and sample homogeneity, grind the mixture thoroughly in an agate mortar. Transfer the ground mixture quantitatively to an Al sample-holder. Clip the holder in the space provided in the diffractometer of an XRD unit and scan the sample from 2° 2θ to 30° 2θ or higher.

16.4.1 Quantitative Interpretations of Diffractograms

For the theory and derivation of the formulas or equations needed in the calculations, reference is made to Chung (1974[a]). The working equation for the quantitative assessment of multi-component systems is:

$$X_i = (X_c/k_i)(I_i/I_c)$$

in which X_i = amount of component or mineral i (in %), X_c = weight fraction of corundum (in %), k_i = reference intensity ratio, I_i = strongest peak intensity of component i, and I_c = strongest peak intensity of corundum.

The reference intensity ratio, k_i, is defined as:

$$k_i = I_i/I_c$$

The symbols in this equation have been defined above.

CHAPTER 17

DIFFERENTIAL THERMAL ANALYSIS

17.1 DEFINITION AND PRINCIPLES

Differential thermal analysis belongs to a group of analytical techniques called *thermal analysis* (Tan et al., 1986). Thermal analysis covers a large number of methods that determine some physical parameter, such as energy, weight, dimension, and evolved volatiles, as a dynamic function of temperature. Among the many methods, three types of thermal analysis find application in soil research, e.g., differential thermal analysis (DTA), differential scanning calorimetry (DSC), and thermogravimetry (TG). Of the three, differential thermal analysis is the most important and widely used technique in the qualitative and quantitative determination of clay minerals (Mackenzie, 1970). However, today DSC instruments are available that can be used for both DTA and DSC analysis. The most modern version is the *modulated DSC (MDSC)*. Differential thermal analysis was developed by Le Chatelier in 1887, and its usefulness was ignored for some time. However, recently it has been employed in geology, in the ceramic, glass, polymer, cement and plaster industries, and in research in chemistry and catalysis. Currently DTA is also used in studies of organic matter, explosives, and radioisotopes. In soil research, it was first employed by Matjeka (1922) for the determination of kaolinite. Russell and Haddock (1940) and Hendricks and Alexander (1940) were perhaps among the first American scientists to recognize it as an important method in analysis of clay minerals. A review of developments in DTA during 1940 to 1960 was published by Murphy (1962). Much of the literature discusses thermal analysis of clays from clay deposits rather than of clays from soils. Indications are currently available that soil clays differ from clay deposits in many respects.

In principle, DTA determines energy changes between a sample and a reference material as the two are heated at a controlled rate. The sample and reference material are heated side by side from 0 to 1000°C. The sample temperature is continuously measured and compared with the temperature of the reference material by means of a set of thermocouples. The heating procedure may cause the sample to undergo a transformation at a certain temperature. For example, heating causes a transformation of α-quartz into β-quartz. This transformation takes place at 573°C and consists of structural changes. Such kind of thermal reaction causes the temperature of the sample to become different from that of the reference material. This difference in temperature is measured and recorded as a function of temperature at which the difference occurs (Figure 17.1). If the temperature of the sample falls below that of the reference material (ΔT is negative) an endothermic peak develops. When the temperature of the sample rises above that of the reference material (ΔT is positive), an exothermic peak develops. When during the heating process no differences in temperature are recorded, a straight line is produced. This portion of the curve or thermogram for which ΔT is approximately zero is considered to be the base line.

The thermal reactions contributing to the change in temperature of the sample are dehydration, dehydroxylation,

Differential Thermal Analysis

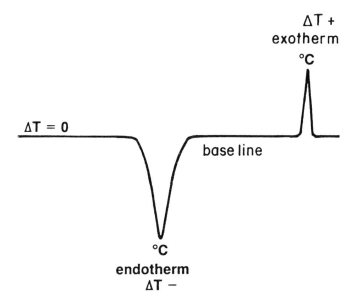

Figure 17.1 Idealized DTA thermogram.

decomposition, boiling, sublimation, vaporization, oxidation, fusion, destruction of crystal structure, and crystalline structure inversion. Dehydration, dehydroxylation, fusion, evaporation and sublimation yield endothermic reactions. On the other hand, oxidation, crystalline structure formation, and some decomposition reactions yield exothermic reactions.

The DTA curves are frequently called *thermograms*. The points on the DTA thermogram, that indicate maximum temperature differences are called *peak temperatures*. They are, of course, situated within the *thermal breaks* (regions in the curves that show the differences in temperature). The size, shape, and temperature of the thermal breaks and peaks are affected not only by the heat of reaction but also by various instrumental factors such as rate of heating; nature of thermocouples; size, shape, and nature of sample holder; position of the thermocouple in the sample hole; nature of the recording instrument; and other details of the instruments itself (Arens, 1951; Mackenzie, 1969).

The temperature differences discussed above are records of net heat or enthalpy changes, and they do not exclude the possibility that both exothermic and endothermic reactions are occurring simultaneously. Hence the absence of a peak (break or loop) in a DTA curve does not necessarily indicate the absence of a thermal reaction. It is conceivable that endothermic and exothermic reactions with equal heat changes might occur simultaneously, canceling each other. Such a cancellation is thought to occur in DTA curves of vermiculite in the temperature region between 700 and 900°C, and in some forms of chlorites in the temperature region between 700 and 800°C (Martin, 1955).

17.2 INSTRUMENTATION

The DTA instrument can be very simple or very complex. For detailed descriptions of commercially available DTA instruments see Tan et al. (1986) and Mackenzie (1970). All of the instruments have in common the following basic components: (a) a furnace or heating unit, (b) a specimen holder, and (c) a temperature regulating and measuring system.

Different kinds of furnaces, operable from 0 to 2800°C, have been used in DTA, with heating furnished by infrared radiation, by high frequency rf oscillation, by a coiled tubing through which heated liquid or gas is circulated, or by resistance elements. Furnaces heated by resistance elements such as the Hoskins furnace are the most common.

The sample holders consist of either a rectangular or circular nickel block or of a ceramic block. They are equipped with cylindrical holes, called *well-holders*, in which the samples and the thermocouples are placed, or they are designed to hold ring-type thermocouples upon which a small platinum pan can be seated. In case of well-holders, it is important that the thermocouples be placed exactly in the center of the holes. The most common thermocouple elements are made of platinum, or a platinum (90%)-rhodium (10%) alloy, or of chromel-P and alumel. The Pt-type thermocouples are preferable in soil analysis, because they are more resistant to corrosion and last longer. These thermocouples are connected in series and measure the differences in heat between the sample and reference material according to the thermal equivalent of Ohm's law:

$$\Delta H = \Delta T/R$$

in which ΔH = difference in heat between sample and reference material, ΔT = measured temperature difference, and R = thermal resistance of the thermocouples or cell.

The size of the two thermocouples in the DTA circuit must be identical; otherwise the DTA thermograms tend to drift upward or downward from the baseline. The thermocouples are placed in the hole either from the side or from the bottom of the sample holder. A schematic diagram of a sample holder cell or unit in a more modern instrument is shown in Figure 17.2.

The recording of temperature differences and furnace temperature is accomplished by placing in the thermocouple circuit either a reflecting galvanometer, which records the variation in emf on photographic paper fastened to a rotating drum, or a pen-and-ink recorder. In older instruments, the temperature differences and the furnace temperature are recorded separately.

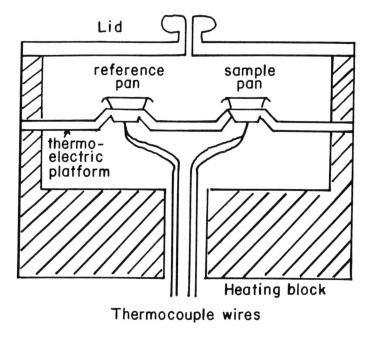

Figure 17.2 Schematic diagram of a sample holder.

17.3 DTA OF SOIL MINERALS AND ORGANIC MATTER

17.3.1 Sample Preparation

Soil Samples Generally, DTA can be conducted on any soil sample. However, soil is complex, because it is a mixture of mineral and organic matter, varying widely in particle size and in physical, chemical and biological properties. Soils also vary in mineralogical composition. All of these factors may influence the result of DTA. The presence of organic matter may interfere with the development of the DTA peaks of soil minerals. Particle size and degree of crystallinity have a pronounced influence on size, shape, and temperature of endothermic peaks. Both the inorganic and organic fractions can be used for analysis by DTA.

Whole Soils The use of whole soils in DTA is sometimes suggested, since this is the medium to which plants and management practices react. DTA curves of whole soils (Figure 17.3) indicate that organic matter caused a strong exothermic reaction culminating into a peak at approximately 300°C, which obscured any endothermic peak that may have developed between 250° and 400°C. In soils with low organic matter content, the latter is not a problem (Figure 17.3, no.2). Removal of organic matter with H_2O_2 or analysis in an inert N_2 atmosphere increases the intensity of endothermic peaks at 270°, 490°, and 573°C for gibbsite, kaolinite, and quartz, respectively.

When whole soils are analyzed, the < 2-mm fraction should be treated with 30% H_2O_2, washed with distilled water, dried and ground again to pass a 2-mm sieve. In general analysis of whole soils gives peaks of low intensities. The peaks may also be obscured because of dilution by high quartz content, charac-

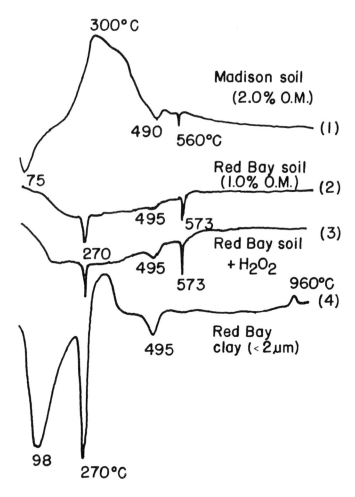

Figure 17.3 DTA thermograms of whole soil and soil clay from Georgia ultisols: (1) and (2) nontreated soil, (3) soil pretreated with H_2O_2, and (4) clay fraction (<2μm) of Red Bay soil.

Differential Thermal Analysis

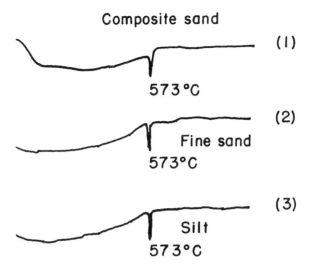

Figure 17.4 DTA thermograms of composite sand, fine sand, and silt fractions separated from a Cecil (Ultisol) AB horizon.

teristic of many soils in the United States. These same peaks are very large and intense if only the clay fractions are analyzed (Figure 17.3, no.4); however, the quartz peak is absent.

Sand Fraction For use in DTA, the composite sand fraction (2.0 - 0.05 mm) can be used directly, or it can be separated first into very coarse sand (2.0 - 1.00 mm), coarse sand (1.00 - 0.50 mm), medium sand (0.50 - 0.25 mm), fine sand (0.25 - 0.10 mm), and very fine sand (0.10 - 0.05 mm). If small sample sizes are to be used the finer fractions are more suitable. When a relatively large amount of sample can be used, as is the case with DTA instruments equipped with well-type sample holders, it is immaterial whether coarse or fine sand is used. As can be noticed from Figure 17.4 (no.1 and 2), the DTA thermogram of the composite sand is identical to that of the fine sand. In both cases a strong endothermal peak of quartz at 573°C is the only characteristic of the curve. Analysis of the sand fraction by DTA is only of importance in the investigation of primary minerals and/or Fe-Mn concretions. In soils of the southern United States and in other parts of the continental United States, quartz is the dominant component of the sand fraction. DTA of the sand fraction of these soils usually results in featureless DTA thermograms, except for the quartz peak.

Silt Fraction The silt fraction of soils (0.05 - 0.002 mm) can be analyzed directly and usually yields thermograms showing more complexity than that of the sand fraction (Figure 17.4, no.3). Often the amount of sesquioxides and kaolinite in the silt size fraction is large enough to yield detectable endothermic peaks at 270 and 498°C, respectively, in addition to the quartz peak at 573°C. Generally the DTA thermogram of silt resembles that of whole soil.

Clay Fraction The clay fraction of soils (< 2µm) can be used directly or can be separated into coarse clay (2.0 - 0.20µm), and fine clay (< 0.2µm) fractions. The choice of the size fraction to be used depends upon many factors, such as the purpose and objective of the study, the desired precision, and the type and quantity of minerals present. A number of investigators have suggested the use of clay separated into various fractions (Jackson, 1956). For the analysis of amorphous minerals, such as allophane, the use of fine clay is preferred. However, for general purposes, the whole clay fraction (< 2µm) gives normally

Differential Thermal Analysis

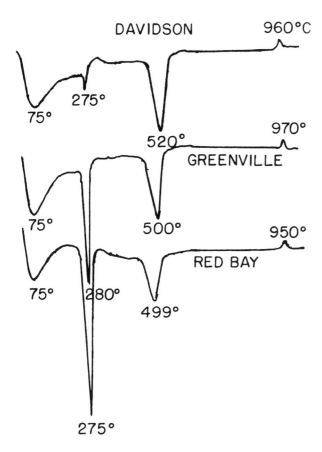

Figure 17.5 DTA thermograms of clay (< 2μm) fractions extracted from a Davidson soil, Greenville soil, and Red Bay soil (Rhodic Paleudults).

satisfactory results for qualitative and quantitative DTA analysis (Figure 17.5).

Soil Organic Fraction   The application of DTA in the analysis of soil organic matter has received increasing research attention with the increased knowledge on extraction and purification of humic matter from soils. The humic fractions isolated can be used directly in DTA, or they can be saturated first with different kinds of cations before analysis (Tan, 1978; Schnitzer and Kodama, 1972). Figure 17.6 indicates that DTA is able to distinguish between humic and fulvic acid. Humic acid is characterized by a DTA curve with a strong exothermic reaction at 400°C, whereas fulvic acid exhibits a curve with an exothermic peak at 500°C. The main decomposition peak of humic acid at 400°C may shift to lower or higher temperatures depending on the different cations used for saturation. The cations Ca^{2+}, Ba^{2+}, Fe^{3+}, Cu^{2+}, and Mn^{2+} usually decrease the thermal stability of humic acid. Indications have also been obtained by the author that different kinds of soils may contain humic acids with different DTA features.

Cation Saturation Saturation of the sample with a known cation prior to DTA has been proposed for clay samples. Clay in soils is expected to be saturated with a variety of cations (Na^+, K^+, Ca^{2+}, Mg^{2+}, and Al^{3+}). Hydration properties of these cations may differ considerably, and will affect the results of DTA analysis. Mackenzie (1970) recommended that for comparative work each sample should receive identical pretreatments. The current author has found that Ca^{2+} saturation of the samples is usually satisfactory. DTA thermograms of kaolinite samples saturated with different cations are shown in Figure 17.7. It appears that the different cations affect both the size and shape of endothermic and exothermic peaks. However, no change in peak temperature is noticed. Ca-kaolinite has the lowest peak intensity at both the 530°C and 1000°C peaks. Peak intensity increases from Na-kaolinite to Al-Kaolinite. Saturation with Al^{3+} produces a very sharp and slender endothermic peak at 530°C and yields in addition an endothermic peak at 125°C. The latter suggests that Al-kaolinite retains considerable amounts of water

Differential Thermal Analysis

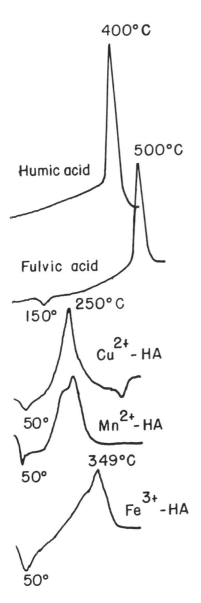

Figure 17.6 DTA thermograms of soil humic fractions and metal-humic acid complexes.

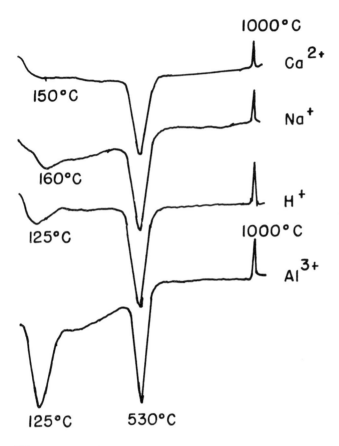

Figure 17.7 DTA thermograms of kaolinite (no.2, Birch Pit, Macon, GA) saturated with Ca^{2+}, Na^+, H^+, and Al^{3+} ions, respectively.

even after overnight storage in a desiccator over $CaCl_2$.

Saturation of smectite with H^+ ions yields curves with a broad peak at 675°C and an S-shape curve at 950°C (Figure 17.8). Ca-smectite exhibits a thermogram with its main

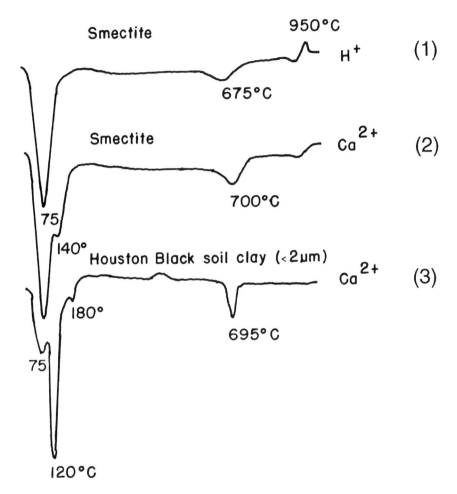

Figure 17.8 DTA thermograms of smectite (montmorillonite, Oklahoma Geological Survey): (1) saturated with H^+ and (2) Ca^{2+} ions; (3) Ca-clay (<2 μm) from a Houston Black soil (Vertisol).

endothermic peak shifted to 700°C (Figure 17.8, no.2). The presence of two peaks at 75° and 140°C for Ca-smectite is in agreement with reports in the literature for smectite treated with Ca^{2+} (Mackenzie, 1970). Thermogram No.3 of Figure 17.8 shows DTA features of a soil smectite from a Houston Black soil (vertisol) for comparison.

Hydration and Solvation   As discussed in the preceding section, hydration of samples prior to DTA may result in changes in the low temperature endothermic peaks at 0 to 200°C. Soil colloids are very reactive due to a large specific surface and electric charge, and may adsorb considerable amounts of water. Moreover, the various types of soil colloids are known to have different capacities for adsorption of moisture. Differences in these factors may result in different DTA curve features. Samples should be equilibrated at a constant humidity to insure that amorphous material and 2:1 lattice-type clays exhibit low temperature endothermic curves that can be used for meaningful comparisons (Jackson, 1956). Equilibration is usually carried out in a desiccator over a compound developing a stable relative humidity. Compounds, such as $Mg(NO_3)_2 \cdot 6H_2O$, $Mg(NO_3)_2$, and H_2SO_4 have been used. The authors have found that keeping samples overnight over $CaCl_2$ produces satisfactory results and will not lead to disappearance of characteristic low temperature endothermic peaks. Figure 17.9 shows DTA curves of Ca-kaolinite kept overnight over $CaCl_2$ and kept open in contact with air in the laboratory (air-dry). The curve of the $CaCl_2$-equilibrated Ca-kaolinite shows practically no low temperature reaction, which is, of course, normal. However, air-dry Ca-kaolinite exhibits a DTA curve with a strong endothermic peak at 125°C. Keeping samples for prolonged periods over $CaCl_2$ may, however, prove to be a disadvantage.

Differential Thermal Analysis

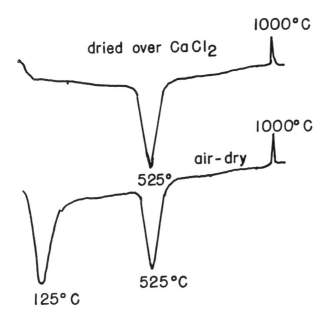

Figure 17.9 DTA thermograms of Ca-kaolinite, dried overnight over $CaCl_2$, and air-dried Ca-kaolinite.

17.3.2 Sample Size

The size of sample to be analyzed is in many cases dictated by the type of DTA instrument available. Most instruments, especially those equipped with well-type sample holders, require sizeable amounts of sample (> 100 mg) to fill the hole. Analysis of a large amount of sample yields extremely large peaks, which are difficult to record on a normal piece of paper. In this case, dilution, by mixing the sample with an inactive material, becomes beneficial. Dilution is also suggested to reduce base-

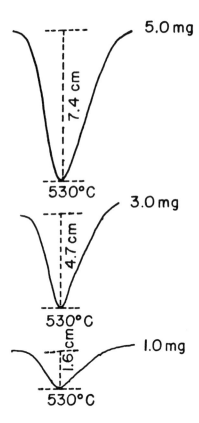

Figure 17.10 The effect of sample size on the main endothermic peak of kaolinite (Mesa Alta, NM).

line drift and/or increase accuracy. For the merits of and objections to dilution, reference is made to Mackenzie (1970).

Instruments, such as the Tracor-Stone apparatus equipped with ring-type thermocouples, or the Dupont 990 Thermal Analyzer (plate-type, Al-pans), require very small samples. The

TA Instruments, Inc. (New Castle, DE) MDSC instrument uses 10-20 mg of sample. Although the amount required for an analysis will vary according to type and thermal characteristics of the material, the size is usually in the order of 1 to 10 mg; as little as 0.5 mg of sample can be used, especially in the analysis of soil humic compounds. These small samples approach the ideal postulated by Mackenzie (1970); i.e., the ideal sample should be an infinitely small sphere surrounding the thermocouple junction.

In qualitative analysis, it is not necessary to weigh the sample for DTA, although comparison of thermograms should be made with those obtained from identical amounts of samples. However, in quantitative DTA analysis, the sample must be weighed accurately. Figure 17.10 shows the effect of size of sample on the main endothermic peak of kaolinite.

17.3.3 Sample Packing

Packing the sample is required when instruments with well-type sample holders are used. A number of methods have been proposed, including hand tapping the sample or the block, and layer packing (Arens, 1951; Tan, 1982). In the latter, the sample is packed around the thermocouple junction as a sandwich between two layers of reference material. The sandwich method yield peaks of approximately one-half the intensity of those obtained by filling the whole cavity with the sample.

Reproducibility in packing is of importance, since differences in packing may create differences in sample density, leading in turn to differences in heat conductivity and thermal diffusivity of the sample. The latter is a problem in the low temperature ranges, where heat transfer is mainly controlled by conduction. In the high temperature region, the effects of packing are less apparent, because heat transfer is mostly through radiation. Both reference and sample materials should be of similar aggregate size and should be packed in a similar bulk density to avoid base-line drift. In plate or ring-type instruments, where a small amount of sample is placed in an Al or Pt cup, packing effects are negligible. Minimum zero drift is usually attained by carefully balancing the relative amounts of reference and/or test sample in the cups.

17.3.4 Furnace Atmosphere

Some control of furnace atmosphere is desirable and often necessary. The furnace atmosphere affects DTA through one or a combination of the following reactions (Schultze, 1969): (1) a change in furnace atmosphere creates a change in partial pressure of the gas atmosphere inducing a shift of peak temperature, (2) interactions occurring between gas used and gaseous products of the sample may change the furnace atmosphere and obscure the appearance of characteristic DTA peaks, or (3) peaks are enhanced because oxidation reactions can be eliminated. A number of methods, such as use of vacuum, static air, static gas, dynamic gas, gas flow over the sample, gas flow through the sample, and self-generated gas have been tested. For the advantages and disadvantages of one method over the other, reference is made to Mackenzie (1970).

When the furnace is evacuated or when an atmosphere of N_2 is introduced during an analysis, the DTA curve of a sample containing organic matter does not have the exothermic break associated with its oxidation; hence for qualitative DTA the elimination of organic matter is unnecessary.

The use of a furnace similar to that employed for total carbon analysis, but modified to the extent that the gases evolved during heating can be trapped at various temperature ranges, enables measurement of the actual losses of water and CO_2 during DTA. The modification consists of attaching to the outlet of the furnace pyrex glass tubes arranged in series for trapping quantitatively the water and CO_2 evolved during heating.

Differential Thermal Analysis

For most analysis, it is sufficient to run DTA in the presence of a self-generated gas atmosphere. However, it is necessary to place a close- but loose-fitting lid on top of the sample holder to maintain uniform pressure and thermal conditions around the sample and reference material. By using a lid, direct radiation of both specimens by the furnace will also be avoided.

17.3.5 Heating Rate

The heating rate is defined as the rate of temperature increase, expressed in degrees centigrade per minute. Correspondingly, the cooling rate is the rate of temperature decrease in degrees centigrade per minute. The heating or cooling rate is constant when the temperature/time curve is linear.

The heating must be controlled at a uniform and steady rate through the analysis. Heating rates used vary from 0.1 to 200°C per minute. For most purposes, heating rates of 5 to 20°/m are used. Although many investigators have stressed the fact that heating rates affect peak height, peak width, peak area, and peak temperature, the present author has found that heating rate is seldom a serious problem. The differences in DTA features obtained by different heating rates (Figure 17.11) indicate that peak height increases gradually and consistently, when heating rates are increased from 15 to 20, 30 and 40°C/m. The most significant difference is obtained between heating at 15°C and 40°C/m. Kaolinite heated at a rate of 15°C per minute yields a broad and relatively shallow endothermic peak at 520°C. When heated at a rate of 40°C/m, a higher and slender endothermic peak was obtained, with the peak temperature shifted to 540°C. The peak temperatures, at 520° or 540°C, are well within the range of kaolinite. Since the shape and size of endothermic peaks obtained by heating at 15 or 40°C/m are well within the detection limits, the choice of heating rate between 15 and 40°C/m makes little difference.

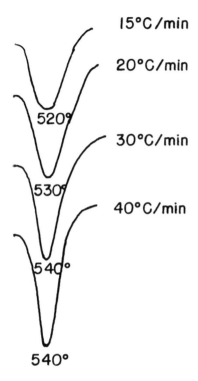

Figure 17.11 The effect of heating rates on the size of the main endothermic peak of kaolinite (Mesa Alta, NM).

17.4 QUALITATIVE INTERPRETATION OF DTA OF SOIL MINERALS

The qualitative identification of soil minerals can be achieved by using the DTA thermograms as fingerprints and comparing them with DTA curves of standard minerals or well-known established soil minerals. Although within a mineral species the DTA features may vary somewhat according to origin, each mineral exhibits thermal reactions, as reflected by endothermic and exothermic peaks, within specific or well-defined limits (Figure 17.12).

Kaolinite DTA curves of kaolinite are generally characterized by a strong endothermic peak between 450 to 600°C, and by a strong exothermic peak between 900 to 1000°C. The endothermic peak is caused by dehydroxylation, whereas the exothermic peak is attributed to formation of γ-alumina and/or mullite.

Halloysite The curve of halloysite is similar to that of kaolinite. However, it has in addition a low temperature (100 - 200°C) endothermic peak of medium to strong intensity for loss of adsorbed water. Some investigators are of the opinion that the shape of the main endothermic peak between 450 and 600°C can be used to distinguish kaolinite from halloysite. This endothermic peak is symmetrical for kaolinite, but asymmetrical for halloysite. However, the present author has noted cases in which endothermic peaks of kaolinite were also asymmetrical.

Smectite This mineral exhibits a DTA curve characterized by a low temperature (100 - 200°C) endothermic peak, an endothermic peak between 600 and 700°C, followed by a weak exothermic peak between 900 and 1000°C.

Vermiculite The DTA curve (not shown) of this mineral resembles closely that of smectite, except for a stronger intensity of the endothermic break at 800 to 900°C and the absence of the main endothermic reaction between 600 and 700°C.

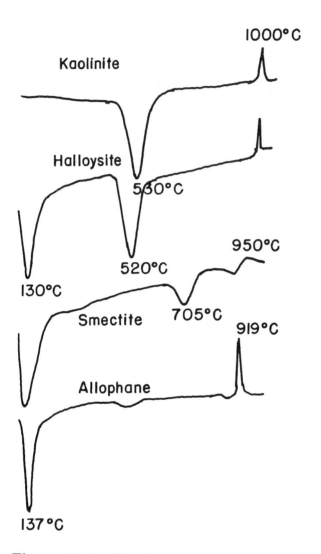

Figure 17.12 DTA thermograms of kaolinite (Macon, GA), halloysite (Bedford, IN), smectite (Oklahoma Geological Survey), and allophane (extracted from an andept).

Differential Thermal Analysis

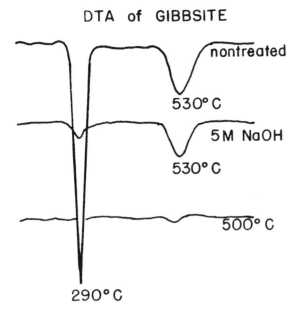

Figure 17.13 Identification of gibbsite using differential dissolution and DTA. The endothermic peak at 530°C indicates the presence of kaolinite.

Allophane Allophane exhibits DTA features with a strong low temperature (50° - 150°C) endothermic peak and a strong exothermic peak between 900 and 1000°C.

The low temperature endothermic reaction is attributed to loss of adsorbed water, whereas the main exothermic peak is caused by γ-alumina formation.

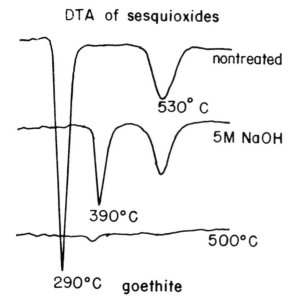

Figure 17.14 Identification of goethite using differential dissolution and DTA.

Gibbsite and Goethite These minerals are usually characterized by DTA curves with a strong endothermic peak only between 290 and 350°C (Figures 17.13 and 17.14). Often goethite and the other iron oxide minerals have their endothermic peaks at higher temperature ranges than gibbsite. Differential dissolution is frequently applied to distinguish gibbsite from goethite. The mineral mixture is shaken with 0.5 M NaOH for 4 h prior to DTA. This treatment will dissolve the gibbsite component (Figure 17.13), leaving goethite unaffected

Differential Thermal Analysis

by analysis with DTA (Figure 17.14).

A list of characteristic endothermic and exothermic peaks of major clay minerals is provided in Table 17.1.

Table 17.1 DTA Endothermic and Exothermic Peak Temperatures of Major Clay Minerals (Tan et al., 1982)

Mineral	Endotherm. peak temp.	Main reaction	Exotherm. peak temp.	Main reaction
	°C		°C	
Kaolinite	500-600	Dehydroxylation	900-1000	γ-alumina formation
Dickite	500-700	Dehydroxylation	900-1000	γ-alumina formation
Nacrite	500-700	Dehydroxylation	900-1000	γ-alumina formation
Halloysite	100-200	Loss of adsorbed water	900-1000	γ-alumina formation
	500-600	Dehydroxylation		
Smectite	100-250	Loss of adsorbed water	900-1000	Recrystallization
	600-750	Dehydroxylation		
Beidellite	100-250	Loss of adsorbed water	900-1000	Recrystallization
	500-600	Dehydroxylation		
Nontronite	100-200	Loss of adsorbed water	900-1000	Recrystallization
	500	Dehydroxylation		

Table 17.1 (Continued)

Mineral	Endotherm. peak temp. °C	Main reaction	Exotherm. peak temp. °C	Main reaction
Vermiculite	150	Loss of adsorbed water	800-900	Recrystallization
	850	Dehydroxylation		
Illite	100-200	Loss of adsorbed water	920-950	Recrystallization
	600	Dehydroxylation		
	900-920	Dehydroxylation		
Chlorite	500-600	Dehydroxylation	800	
Gibbsite	250-350	Dehydroxylation		
Boehmite	570	Dehydroxylation		
Diaspore	400-500	Dehydroxylation		
Goethite	300-400	Dehydroxylation		
Limonite	300	Dehydroxylation		
Magnetite	530	Fe(II) → Fe(III)		
Ilmenite	525-700	Fe(II) → Fe(III)		
Quartz	573	α to β inversion		
Allophane	50-150	Loss of adsorbed water	800-900	γ-alumina formation
Muscovite	500-800			

Table 17.1 (Continued)

Mineral	Endotherm. peak temp.	Main reaction	Exotherm. peak temp.	Main reaction
	°C			
Calcite	970	CaO formation		
Dolomite	740	MgO formation		
Aragonite	447	CaO formation		

17.5. QUALITATIVE INTERPRETATION OF DTA OF ORGANIC MATTER

Not much is known in this respect. From the investigations of the current author, it can be observed that DTA is able to distinguish between humic acid and fulvic acid (Figure 17.6). Indications have been obtained, that different kinds of soils may contain humic acids with different kinds of DTA features.

Humic Acid Humic acid exhibits a DTA curve with a strong exothermic peak at 400°C to 410°C (Tan, 1978). However, Dupuis (1971) reported that humic acid is characterized by a small and strong exothermic peak at 340° and 430°C, respectively. There is a possibility, that the humic acid from Dupuis, extracted from French soils, is slightly different in nature than the humic acid obtained by Tan from southeastern U.S. soils.

Fulvic Acid The DTA curve of fulvic acid exhibits a strong exothermic peak at 500°C. This higher peak temperature indicates that fulvic acid is thermally slightly more stable than humic acid.

Humic Acid-Metal Chelates The main decomposition peak of humic acid at 400°C shifts to lower or higher temperatures after saturation, complex formation, or chelation with cations. The cations, Ca^{2+}, Ba^{2+}, Fe^{3+}, Cu^{2+}, and Mn^{2+}, usually shift this endothermic peak to lower temperatures. This decrease in decomposition temperature means that the thermal stability of humic acid is decreased by interactions with the cations stated above. Schnitzer and Kodama (1972) believe that monovalent cations yield relatively thermally stable metal-humic acid chelates, since they react only with the carboxyl groups of the HA molecule. On the other hand, di- and trivalent cations are interacting with both the carboxyl and phenolic-OH groups of the humic molecule, causing a severe strain on the molecular structure. This renders the formed metal-HA compound less stable against thermal decomposition. Dupuis (1971) indicates that the degree of thermal stability of the metal-HA chelates also depends on the cation saturation of the humic molecule. At a low ratio of metal to HA, the complex formed is pseudo-soluble, and exhibits a high resistance against thermal decomposition. However, at a high ratio, the metal-HA chelate becomes insoluble, and is thermally less stable than the pseudo-soluble chelates.

17.6 QUANTITATIVE INTERPRETATION OF DTA OF SOIL MINERALS

In qualitative analysis, as discussed in the preceding sections, it is not necessary to weigh the sample for DTA, although comparison of curves should be made with curves obtained from identical amounts of samples. However, in quantitative DTA analysis, the sample must be weighed accurately, and a standard curve must be prepared for quantitative deter-

Differential Thermal Analysis

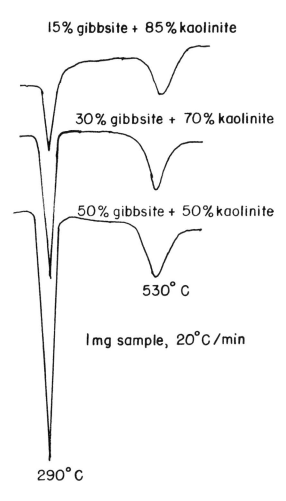

Figure 17.15 DTA thermograms of gibbsite and kaolinite mixtures.

mination. The limitation in quantitative analysis by DTA is the size of the main endothermic or exothermic peak. Only minerals that produce sizable and well-delineated peaks can be analyzed quantitatively, e.g., gibbsite and kaolinite.

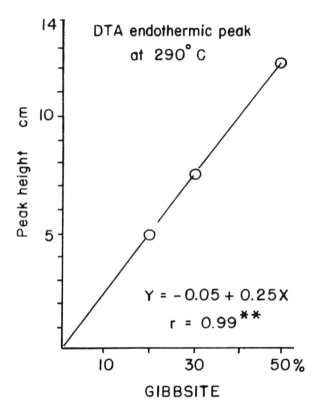

Figure 17.16 Relationship between the heights of the endothermic peak of gibbsite and the gibbsite content. The correlation is significant at the 99% level of probability.

17.6.1 Preparation of a Standard Curve

Prepare mixtures of 15% gibbsite + 85 % kaolinite, 30% gibbsite + 70% kaolinite, and 50% gibbsite + 50% kaolinite. Weigh 10 mg of each mixture separately in Pt cups and ana-

Differential Thermal Analysis

lyze them by DTA. Larger amounts of samples are required for some instruments. However, with a Stone-Tracor DTA instrument equipped with an automatic program controller, 1 mg of sample is usually sufficient.

The results (Figure 17.15) show that the height (or area) of the main endothermic peak of gibbsite at 290°C increases proportionally with increased gibbsite content in the mixture. The regression is linear, and the correlation between the heights of the endothermic peaks and the corresponding gibbsite concentrations is significant at the 99% level of probability (Figure 17.16).

The *unknown* clay sample is then analyzed by DTA as discussed above, and its gibbsite concentration is determined by peak height from the standard curve in Figure 17.16.

APPENDIX A

Fundamental Constants

Symbol	Name	Value
c	Velocity of light	2.9979×10^{10} cm/sec
e	Electronic charge	1.6021×10^{-19} C
L	Avogadro's number	6.0225×10^{23}
h	Max Planck's constant	6.6256×10^{-27} erg sec
F	Faraday constant	96.487 C/Eq
R	Gas constant	82.056 cm^3 atm/(mol)(deg) 1.9872 cal/(mol(deg) 8.3143 J/(mol(deg)
K	Kelvin temperature	-273°C = 0 K
k	Boltzmann constant	1.3805×10^{-16} erg/deg

APPENDIX B

Greek Alphabet

Greek letter	Greek name	Greek letter	Greek name
A α	Alpha	N υ	Nu
B ß	Beta	Ξ ξ	Xi
Γ γ	Gamma	O o	Omicron
Δ δ	Delta	λ π	Pi
E ε	Epsilon	P ρ	Rho
Z ζ	Zeta	Σ σ	Sigma
H η	Eta	T τ	Tau
Θ θ	Theta	Υ φ	Upsilon
I ι	Iota	Φ φ	Phi
K κ	Kappa	X χ	Chi
Λ λ	Lambda	Ψ ψ	Psi
M μ	Mu	Ω ω	Omega

APPENDIX C

Atomic Weights of the Major Elements in Soils

Element	Symbol	Atomic number	Atomic weight
Aluminum	Al	13	027.0
Antimony	Sb	51	121.8
Argon	Ar	18	039.9
Arsenic	As	33	074.9
Barium	Ba	56	137.3
Beryllium	Be	04	009.0
Bismuth	Bi	83	209.0
Boron	B	05	010.8
Bromine	Br	35	079.9
Calcium	Ca	20	040.1
Carbon	C	06	012.0
Cesium	Cs	55	132.9
Chlorine	Cl	17	035.5
Chromium	Cr	24	052.0
Cobalt	Co	27	058.9
Copper	Cu	29	063.5
Fluorine	F	09	019.0
Gallium	Ga	31	069.7
Germanium	Ge	32	072.6
Gold	Au	79	197.0
Helium	He	02	004.0
Hydrogen	H	01	001.0
Iodine	I	53	126.9
Iridium	Ir	77	192.2

Appendix C

Atomic Weights of the Major Elements in Soils

Element	Symbol	Atomic number	Atomic weight
Iron	Fe	26	055.9
Krypton	Kr	36	083.8
Lanthanum	La	57	138.9
Lead	Pb	82	207.2
Lithium	Li	03	006.9
Magnesium	Mg	12	024.2
Manganese	Mn	25	054.9
Mercury	Hg	80	200.6
Molybdenum	Mo	02	095.9
Nickel	Ni	28	058.7
Nitrogen	N	07	014.0
Oxygen	O	08	016.0
Phosphorus	P	15	031.0
Platinum	Pt	78	195.1
Potassium	K	19	039.1
Radon	Rn	86	222.0
Radium	Ra	88	226.1
Rhodium	Rh	45	102.9
Rubidium	Rb	37	085.5
Selenium	Se	34	079.0
Silicon	Si	14	028.1
Silver	Ag	47	107.9
Sodium	Na	11	023.0
Strontium	Sr	38	087.6
Sulfur	S	16	032.1
Tantalum	Ta	03	180.9
Tellurium	Te	52	127.6
Thallium	Tl	81	204.4
Thorium	Th	90	232.1
Tin	Sn	50	118.7
Titanium	Ti	22	047.9

Appendix C

Atomic Weights of the Major Elements in Soils

Element	Symbol	Atomic number	Atomic weight
Tungsten	W	74	183.9
Uranium	U	92	238.0
Vanadium	V	23	050.9
Xenon	Xe	54	131.3
Yttrium	Y	39	088.9
Zinc	Zn	30	065.4
Zirconium	Zr	40	091.2

APPENDIX D

X-ray Diffraction 2θ d-Spacing Conversion Table

2θ d Spacing Values for Cu Kα Radiation with λ = 1.5405 Å (0.1540 nm)

2θ	0.0	0.1	0.2	0.3	0.4	0.5	0.6	0.7	0.8	0.9
0	∞	882.63	441.32	294.21	220.66	176.53	147.11	126.09	110.33	98.07
1	88.263	80.245	73.555	67.897	63.047	58.845	55.167	51.922	49.038	46.45
2	44.135	42.033	40.122	38.378	36.779	35.308	33.950	32.693	31.526	30.44
3	29.425	28.476	27.587	26.751	25.964	25.223	24.522	23.859	23.232	22.63
4	22.071	21.532	21.020	20.531	20.065	19.619	19.193	18.785	18.394	18.01
5	17.659	17.312	16.979	16.660	16.352	16.054	15.768	15.491	15.225	14.96
6	14.717	14.476	14.243	14.017	13.798	13.586	13.381	13.181	12.988	12.80
7	12.617	12.440	12.267	12.099	11.936	11.777	11.622	11.471	11.325	11.18
8	11.042	10.906	10.773	10.644	10.517	10.394	10.273	10.155	10.040	9.927
9	9.8168	9.7098	9.6042	9.5010	9.4001	9.3015	9.2053	9.1105	9.0173	8.926
10	8.8378	8.7500	8.6645	8.5506	8.4989	8.4181	8.3387	8.2609	8.1847	8.110
11	8.0360	7.9644	7.8935	7.8234	7.7549	7.6880	7.6220	7.5571	7.4932	7.430
12	7.3688	7.3081	7.2484	7.1897	7.1320	7.0751	7.0192	6.9642	6.9100	6.856
13	6.8042	6.7524	6.7015	6.6513	6.6019	6.5532	6.5053	6.4550	6.4114	6.365
14	6.3203	6.2757	6.2317	6.1883	6.1456	6.1035	6.0619	6.0209	5.9804	5.940
15	5.9011	5.8623	5.8239	5.7860	5.7488	5.7119	5.6755	5.6395	5.6041	5.569
16	5.5345	5.5004	5.4666	5.4333	5.4004	5.3679	5.3358	5.3040	5.2727	5.241
17	5.2111	5.1809	5.1510	5.1214	5.0922	5.0633	5.0348	5.0065	4.9787	4.951
18	4.9238	4.8968	4.8701	4.8437	4.8176	4.7918	4.7663	4.7410	4.7160	4.691
19	4.6669	4.6426	4.6187	4.5950	4.5715	4.5482	4.5253	4.5026	4.4801	4.457
20	4.4357	4.4138	4.3922	4.3708	4.3496	4.3287	4.3079	4.2872	4.2669	4.246
21	4.2267	4.2069	4.1872	4.1678	4.1486	4.1295	4.1106	4.0919	4.0733	4.055
22	4.0367	4.0187	4.0008	3.9831	3.9656	3.9481	3.9309	3.9139	3.8969	3.880
23	3.8635	3.8469	3.8306	3.8144	3.7983	3.7824	3.7666	3.7509	3.7354	3.720
24	3.7047	3.6896	3.6746	3.6596	3.6449	3.6302	3.6157	3.6013	3.5870	3.572
25	3.5587	3.5448	3.5309	3.5172	3.5036	3.4901	3.4767	3.4634	3.4502	3.437
26	3.4241	3.4112	3.3984	3.3857	3.3731	3.3606	3.3482	3.3359	3.3236	3.311
27	3.2995	3.2875	3.2758	3.2639	3.2522	3.2406	3.2291	3.2176	3.2063	3.195
28	3.1839	3.1727	3.1617	3.1508	3.1399	3.1291	3.1184	3.1078	3.0973	3.086
29	3.0763	3.0660	3.0557	3.0455	3.0354	3.0253	3.0153	3.0054	2.9955	2.985
30	2.9760	2.9664	2.9568	2.9472	2.9377	2.9283	2.9190	2.9098	2.9005	2.891
31	2.8823	2.8732	2.8643	2.8553	2.8465	2.8376	2.8289	2.8202	2.8116	2.802
32	2.7945	2.7859	2.7775	2.7691	2.7608	2.7526	2.7443	2.7362	2.7281	2.720
33	2.7120	2.7040	2.6961	2.6882	2.6804	2.6727	2.6649	2.6573	2.6496	2.642
34	2.6345	2.6270	2.6195	2.6121	2.6048	2.5974	2.5902	2.5830	2.5757	2.568
35	2.5615	2.5541	2.5474	2.5404	2.5334	2.5295	2.5196	2.5129	2.5060	2.499
36	2.4926	2.4859	2.4793	2.4727	2.4661	2.4596	2.4531	2.4466	2.4402	2.433
37	2.4274	2.4211	2.4149	2.4086	2.4024	2.3962	2.3901	2.3840	2.3779	2.371
38	2.3659	2.3599	2.3540	2.3480	2.3421	2.3362	2.3305	2.3247	2.3189	2.313
39	2.3074	2.3018	2.2962	2.2905	2.2849	2.2794	2.2739	2.2684	2.2629	2.257

APPENDIX E

U.S. Weights and Measures

LAND MEASURE
1 foot	12 inches
1 yard	3 feet
1 mile	5280 feet
1 mile	1760 yards
1 square foot	144 square inches
1 square yard	9 square foot
1 square mile	640 acres
1 acre	4840 square yards
1 acre	43 560 square feet

AVOIRDUPOIS WEIGHT
(Used in weighing all articles except drugs, gold, silver and precious stones)

1 pound, lb	16 ounces
1 hundredweight (cwt)	100 lbs
1 ton	20 cwt
1 ton	2000 lbs
1 long ton	2240 lbs

TROY WEIGHT (Used in weighing gold, silver and precious stones)

1 pound, lb	12 ounces

DRY MEASURE
1 quart	2 pints
1 bushel	32 quarts

U.S. Weights and Measures

LIQUID MEASURE
1 quart	2 pints
1 gallon	4 quarts
1 barrel	31.5 gallons

FLUID MEASURE
1 fluid dram	60 minims
1 fluid ounce	8 fluid drams
1 pint	16 fluid ounce
1 gallon	8 pints

GRAIN WEIGHTS PER BUSHEL
Barley	48 lbs
Beans	60 lbs
Bran	20 lbs
Buckwheat	42-52 lbs
Clover seed	60 lbs
Corn (in the ear husked)	70 lbs
Corn (shelled)	56 lbs
Corn meal	48 lbs
Flax seed	56 lbs
Malt	30-38 lbs
Millet seed	50 lbs
Oats	32 lbs
Peas	60 lbs
Rye	56 lbs
Wheat	60 lbs

APPENDIX F

International System of Units (SI)

Basic unit	Symbol
Ampere (electrical current)	A
Candela (luminous intensity)	cd
Kelvin (thermodynamic temperature)	K
Kilogram (mass)	kg
Meter (length)	m
Mole (amount of substance)	mol
Second (time)	s

Factors for Converting U.S. Units into SI Units

U.S. unit	SI unit	To obtain SI unit multiply U.S. unit by
Acre	Hectare, ha	0.405
Acre	Square meter, m^2	4.05×10^3

Appendix F

Factors for Converting U.S. Units into SI Units

U.S. unit	SI unit	To obtain SI unit multiply U.S. unit by
Ångstrom	Nanometer, nm	10^{-1}
Atmosphere	Megapascal, MPa	0.101
Bar	Megapascal, MPa	10^{-1}
Calorie	Joule, J	4.19
Cubic foot	Liter, L	28.3
Cubic inch	Cubic meter, m^3	1.64×10^{-5}
Curie	Becquerel, Bq	3.7×10^{10}
Degrees, °C (+273, temperature)	Degrees, K	1
Degrees, °F (-32, temperature)	Degrees, °C	0.556
Dyne	Newton, N	10^{-5}
Erg	Joule, J	10^{-7}
Foot	Meter, m	0.305
Gallon	Liter, L	3.78
Gallon per acre	Liter per ha	9.35
Inch	Centimeter, cm	2.54
Micron	micrometer, µm	1
Mile	Kilometer, km	1.61
Mile per hour	Meter per second	0.477
Millimho per cm	decisiemens per m, dS m^{-1}	1
Ounce (weight)	Gram, g	28.4
Ounce (fluid)	Liter, L	2.96×10^{-2}
Pint	Liter, L	0.473
Pound	Gram, g	454
Pound per acre	Kilogram per ha	1.12
Pound per cubic foot	Kilogram per m^3	16.02
Pound per square foot	Pascal, Pa	47.9
Pound per square inch	Pascal, Pa	6.9×10^3
Quart	Liter, L	0.946

Appendix F

Factors for Converting U.S. Units into SI Units

U.S. unit	SI unit	To obtain SI unit multiply U.S. unit by
Square foot	Square meter, m^2	9.29×10^{-2}
Square inch	Square cm, cm^2	6.45
Square mile	Square kilometer, km^2	2.59
Ton (2000 lbs)	Kilogram, kg	907
Ton per acre	Megagram per ha	2.24

REFERENCES AND ADDITIONAL READINGS

Adams, F. 1974. Soil solution. p.441-481. In: *The Plant Root and Its Environment*, E.W. Carlson (ed). Univ. Press of Virginia, Charlottesville, VA.

Adams, F., and C.E. Evans. 1962. A rapid method for measuring lime requirement of Red-Yellow Podzolic soils. Soil Sci. Soc. Am. Proc. 26:355-357.

Adler, H.H., P.F. Kerr, E.E. Bray, N.P. Stevens, J.M. Hunt, W.D. Keller, and E.E. Pickett. 1950. Infrared spectra of reference clay minerals. Amer. Petroleum Inst. project 49. Prelim. report No.8. Columbia Univ., New York, NY.

Allison, L.E. 1960. Wet-combustion apparatus and procedure for organic and inorganic carbon determination in soils. Soil Sci. Soc. Am. Proc. 24:36-40.

Allison, L.E. 1965. Organic carbon. p.1367-1378. In: *Methods of Soil Analysis*, Part 2, C.A. Black (ed-in-chief). Agronomy series No. 9. Am. Soc. Agronomy, Inc., Madison, WI.

Allmares, R.R. 1965. Bias. p.24-42. In *Methods of Soil Analysis*, Part 1, C.A. Black (ed-in-chief). Agronomy series No.9. Am. Soc. Agronomy, Madison, WI.

Anderson, G. 1980. Assessing organic phosphorus in soils. p.411-431. In: *The Role of Phosphorus in Agriculture*, F.E. Kasawneh, E.C. Sample, and E.J. Kamprath (eds). Am. Soc. Agronnomy, Madison, WI.

Arens, P.L. 1951. A study on the differential thermal analysis of clays and clay minerals. Proefschrift Landbouw Hogeschool, Wageningen. Excelsiors drukkerij, s'Gravenhage, The Netherlands.

Armstrong, D., B. Agerton, and S. Martin (eds). 1984. *The Diagnostic Approach. Better Crops*. Official and Tentative Methods of Analysis of the Association of Agricultural Chemists. Washington, D.C.

Association of Official Agricultural Chemists. 1940. *Official and Tentative Methods of Analysis of the Association of Agricultural Chemists*. Published by A.O.A.C., Washington, D.C.

Bailey, E.G. 1909. Accuracy in sampling coal. J. Industrial & Engin. Chem. 1:161-178.

Baker, A.S., and R. Smith. 1969. Extracting solution for potentiometric determination of nitrate in plant tissue. J. Agr. Food Chem. 17:1284-1287.

Baker, D.E. 1971. A new approach to soil testing. Soil Sci. 112:6381-391.

Barnhisel, R., and P.M. Bertsch. 1982. Aluminum. p.275-300. In: *Methods of Soil Analysis*, Part 2, A.L. Page, R.H. Miller, and D.R. Keeney (eds). 2nd ed. Agronomy series No.9. Am. Soc. Agronomy, Inc., Madison, WI.

Beaty, E.R., and K.H. Tan. 1972. Organic matter, N, and base accumulation under Pensacola bahiagrass. J. Range Management 25:38-40.

Bernas, B. 1968. A new method for decomposition and comprehensive analysis of silicates by atomic absorption spectrometry. Anal. Chem. 40:1682-1686.

Berry, C. 1970. Inorganic index to the powder diffraction file. Joint Committee on Powder Diffraction Standards. Pittsburgh, PA.

Blake, G.R., and K.H. Hartge. 1986. Bulk density. p.363-375. In: *Methods of Soil Analysis*, Part 1, A. Klute (ed). Agronomy series No. 9. Am. Soc. Agronomy and Soil Sci. Soc. Am., Inc., Publ. Madison, WI.

Bouyoucos, G.J. 1953. More durable plaster of paris moisture blocks. Soil Sci. 76:447-451.

References

Bouyoucos, G.J. 1962. Hydrometer method improved for making particle size analysis of soils. Agronomy J. 54:464-465.

Bower, C.A., and E. Truog. 1940. Base exchange capacity determination as influenced by the nature of cation employed and formation of basic exchange salt. Soil Sci. Soc. Am. Proc. 5: 86-89.

Brady, N.C. 1990. *The Nature and Properties of Soils*. Macmillan Publ. Co., New York, NY.

Bragg, W.L. 1933. *The Crystalline State*. Macmillan Publ. Co., New York, NY.

Bremner, J.M. 1950. Some observations on the oxidation of soil organic matter in the presence of alkali. J. Soil Sci. 1: 198-204.

Bremner, J.M. 1960. Determination of nitrogen in soil by the Kjeldahl method. J. Agri. Sci. 55:11-33.

Bremner, J.M. 1965. Inorganic forms of nitrogen. p.1179-1237. In: *Methods of Soil Analysis*, Part 2, C.A. Black (ed-in-chief). Agronomy series No.9. Am. Soc. Agronomy, Inc., Publ., Madison, WI.

Bremner, J.M., and C.S. Mulvaney. 1982. Nitrogen-Total. p.595-624. In: *Methods of Soil Analysis*, Part 2, A.L. Page, R.H. Miller, and D.R. Keeney (eds). 2nd ed. Agronomy series No.9. Am. Soc. Agronomy, Inc., Publ., Madison, WI.

Brindley, G.W. 1980. Quantitative x-ray mineral analysis of clays. p.141-438. In: *Crystal Structures of Clay Minerals and their X-ray Identification*. Mineral. Soc. monograph No.5. Mineralogical Society, London.

Bullock, P., N. Fedoroff, A. Jongerius, G. Stoops, and T. Tursina. 1985. *Handbook for Soil Thin Section Description*. Wayne Research Publ., Wolverhampton, England.

Calderoni, G., and M. Schnitzer. 1984. Effects of age on the chemical structure of paleosol humic acids and fulvic acids. Geochim. Cosmochim. Acta 48: 2045-2051.

Cassell, D.K., and A. Klute. 1986. Water potential: Tensiometry. p.563-596. In: *Methods of Soil Analysis*, Part 1, A. Klute (ed).

Agronomy series No.9. Am. Soc. Agronomy and Soil Sci. Soc. Am., Inc., Publ., Madison, WI.

Chapman, H.D., J.H. Axley, and D.S. Curtis. 1940. The determination of pH at soil moisture contents approximating field conditions. Soil Sci. Soc. Am. Proc. 5: 191-200.

Chesnin, L., and C.H. Yien. 1950. Turbidimetric determination of available sulfates. Soil Sci. Soc. Am. Proc. 15:149-151.

Chung, F.H. 1974[a]. Quantitative interpretation of x-ray diffraction patterns of mixtures. I. Matrix-flushing method for quantitative multicomponent analysis. J. Appl. Cryst. 7:519-525.

Chung, F.H. 1974[b]. Quantitative interpretation of x-ray diffraction patterns of mixtures. II. Adiabatic principle of x-ray diffraction analysis of mixtures. J. Appl. Cryst. 7:527-531.

Clark, F.E., and K.H. Tan. 1969. Identification of a polysaccharide ester linkage in humic acid. Soil Biol. Biochem. 1:75-81.

Clark, R.B. 1984. Physiological aspects of calcium, magnesium, and molybdenum deficiencies in plants. p.99-170. In: *Soil Acidity and Liming*, F. Adams (ed). 2nd ed. Agronomy series No.12. Am. Soc. Agronomy, Inc., Publ., Madison, WI.

Clements, H.F. 1960. Crop logging of sugar cane in Hawaii. p.131-147. In: *Plant Analysis and Fertilizer Problems*, W. Reuther (ed). Am. Inst. Biol. Sci., Washington, D.C.

Cline, M.G. 1944. Principles of soil sampling. Soil Sci. 58:275-288.

Cline, M.G. 1945. Methods of collecting and preparing soil samples. Soil Sci. 59: 3-5.

Cochran, W.G. 1963. *Sampling Techniques*. 2nd ed. John Wiley & Sons, New York, NY.

Coghill, W.H., and F.D. Devaney. 1937. Ball-mill grinding. U.S. Bureau Mines. Techn. Paper No.581.

Coleman, E.A., and T.M. Hendrix. 1949. Fiberglass electrical soil-moisture instrument. Soil Sci. 67: 425-438.

Cox, F.R., and E.J. Kamprath. 1973. Micronutrient soil test.

p.289-317. In: *Micronutrient in Agriculture*, J.J. Mortvedt, P.M. Giordano, and W.L. Lindsay (eds). Soil Sci. Soc. Am., Inc., Madison, WI.

Dumas, J.B.A. 1831. Procedes de l'analyse organique. Ann. Chem. Phys. 2,47:198-213.

Dupuis, T. 1971. Caracterisation par analyse thermique differentiele des complexes de l'aluminum avec les acides fulviques et humiques. J. Thermal Anal. 3:281-288.

Farmer, V.C. (ed). 1974. *The Infrared Spectra of Minerals*. Mineral. Soc., London.

Farmer, V.C. 1979. The role of infrared spectroscopy in a soil research institute: Characterization of inorganic materials. Eur. Spectrosc. News 25:25-27.

Ferraro, J.R. (ed). 1982. *The Sadtler Infrared Spectra Handbook of Minerals and Clays*. Sadtler Research Lab., Philadelphia, PA.

Fitts, J.W., and W.L. Nelson. 1956. The determination of lime and fertilizer requirements of soils through chemical test. Advances Agronomy VIII:241-282.

Flaig, W., H. Beutelspacher, and E. Rietz. 1975. Chemical composition and physical properties of humic substances. p.1-21. In: *Soil Components*, Vol.1, *Organic Components*, J.E. Gieseking (ed). Springer-Verlag, New York, NY.

Foy, C.D. 1978. The physiology of metal toxicity in plants. Ann. Rev. Plant Physiol. 29:511-516.

Fujimoto, C.K., and G.D. Sherman. 1945. The effect of drying, heating and wetting on the level of exchangeable manganese in Hawaiian soils. Soil Sci. Soc. Am. Proc. 10:107-112.

Fujimoto, C.K., and G.D. Sherman. 1948. Manganese availability as influenced by steam sterilization of soil. Am. Soc. Agronomy J. 40:527-534.

Gallaher, R.N., and J.B. Jones, Jr. 1976. Total, extractable, and oxalate calcium and other elements in normal and mouse ear pecan tree tissues. J. Am. Soc. Hort. Sci.. 101:692-696.

Gardner, W.H. 1986. Water content. p.493-544. In: *Methods of*

Soil Analysis, Part 1, A. Klute (ed),. Agronomy series No.9. Am. Soc. Agronomy and Soil Sci. Soc. Am., Inc., Publ., Madison, WI.

Gee, G.W., and J.W. Bauder. 1986. Particle-size analysis. p.383-411. In: *Methods of Soil Analysis*, Part 1, A. Klute (ed). Agronomy series No.9. Am. Soc. Agronomy and Soil Sci. Soc. Am., Inc., Publ., Madison, WI.

Golden, L.B., N. Gammon, and R.P. Thomas. 1942. A comparison of methods for determining the exchangeable cations and the exchange capacity of Maryland soils. Soil Sci. Soc. Am. Proc. 7:134-161.

Hallmark, C.T., L.P. Wilding, and N.E. Smeck. 1982. Silicon. p.263-273. In: *Methods of Soil Analysis*, Part 2, A.L. Page, R.H. Miller, and D.R. Keeney (eds). 2nd ed. Agronomy series No.9. Am. Soc. Agronomy, Inc., Madison, WI.

Hansen, M.H., W.N. Hurwitz, and W.G. Madow. 1953. *Sample Survey Methods and Theory*. John Wiley & Sons, New York, NY.

Harmsen, G.W., and G.J. Kolenbrander. 1965. Soil inorganic nitrogen. p.43-92. In: *Soil Nitrogen*, W.V. Bartholomew and F.E. Clark (eds). Agronomy series No.10. Am. Soc. Agronomy, Inc., Publ., Madison, WI.

Hendricks, S.B., and L.T. Alexander. 1940. Semiquantitative estimation of montmorillonite and clays. Soil Sci. Soc. Am. Proc. 5:95-99.

Hesse, P.R. 1972. *A Textbook of Soil Chemical Analysis*. Chem. Publ. Co., Inc., New York, NY.

Hughes, R., and B. Bohor. 1970. Random clay powders prepared by spray drying. Am. Mineral. 55: 1780-1786.

Hurlbut, Jr., C.S., and C. Klein. 1977. *Manual of Mineralogy*. John Wiley & Sons, New York, NY.

Isaac, R.A., and W.C. Johnson. 1976. Determination of total nitrogen in plant tissue. J. Assoc. Off. Anal. Chem. 59:98-100.

Jackson, M.L. 1956. *Soil Chemical Analysis - Advanced Course*. Publ. by the author, Dept. Soils, Univ. of Wisconsin, Madison, WI.

References

Jackson, M.L. 1958. *Soil Chemical Analysis*. Prentice-Hall, Inc., Englewood Cliffs, NJ.

Jackson, M.L. 1960. *Soil Chemical Analysis*. Prentice-Hall, Inc., Englewood Cliffs, NJ.

Jenkins, R., R.W. Gould, and D.A.Gedeke. 1981. *Quantitative X-ray Spectrometry*. Marcel Dekker, Inc., New York, NY.

Jenny, H. 1941. *Factors of Soil Formation*. McGraw-Hill, New York, NY.

Jones, Jr., J.B. 1963. Effect of drying on ion accumulation in corn leaf margins. Agron. J. 55:579-580.

Jones, Jr., J.B. 1970. Distribution of fifteen elements in corn leaves. Comm. Soil Sci. Plant Anal. 1:27-33.

Jones, Jr., J.B., 1976. Elemental analysis of biological substances by direct-reading spark emission spectroscopy. Am. Lab. 8:15-20.

Jones, Jr., J.B. 1983. Soil test works when used right. Part 2. Solutions 27:61-70.

Jones, Jr., J.B. 1985. Soil testing and Plant Analysis: Guides to fertilization of horticultural crops. Hort. Reviews 7:1-68.

Jones, Jr., J.B., and W.W. Budzynski. 1980. A professional seminar on agronomics. Agric. Assoc., Inc., West Lafayette, IN.

Jones, Jr., J.B., J.E. Pallas, and J.R. Stansell. 1980. Tracking the elemental content of leaves and other plant parts of the peanut under irrigated culture in the sandy soils of South Georgia. Comm. Soil Sci. Plant Anal. 11:81-92.

Jones, Jr., J.B., B. Wolf, and H.A. Mills. 1991. *Plant Analysis Hand-book*. Micro-Macro Publ., Inc., Athens, GA.

Kamprath, E.J., and M.E. Watson. 1980. Conventional soil and tissue tests for assessing the phosphorus status of soils. p.433-469. In: *The Role of Phosphorus in Agriculture*, F.E. Khasawneh, E.C. Sample, and E.J. Kamprath (eds). Am. Soc. Agronomy, Madison, WI.

Keeney, D.R., and D.W. Nelson. 1982. Nitrogen - Inorganic forms. p.643-698. In: *Methods of Soil Analysis*, Part 2, A.L.

Page, R.H. Miller, and D.R. Keeney (eds). 2nd ed. Agronomy series No.9. Am. Soc. Agronomy, Soil Sci. Soc. Am., Inc., Publ., Madison, WI.

Kelley, W.P. 1948. *Cation Exchange in Soils.* Reinhold Publ. Corp., New York, NY.

Kempthorne, O., and R.R. Allmaras. 1965. Errors of observation. p.1-23. In: *Methods of Soil Analysis,* Part 1, C.A. Black (ed-in-chief). Agronomy series No.9. Am. Soc. Agronomy, Inc., Madison, WI.

Khasawneh, F.E., E.C. Sample, and E.J. Kamprath (eds). 1980. *The Role of Phosphorus in Agriculture.* Am. Soc. Agronomy, Madison, WI.

Kilmer, V.J., and L.T. Alexander. 1949. Methods of making mechanical analyses of soils. Soil Sci. 68:15-24.

Knudson, D., G.A. Peterson, and P.F. Pratt. 1982. Lithium, sodium, and potassium. p.225-246. In: *Methods of Soil Analysis,* Part 2, A.L. Page, R.H. Miller, and D.R. Keeney (eds). 2nd ed. Agronomy series No.9. Am. Soc. Agronomy, Inc., Publ., Madison, WI.

Kolthoff, I.M., and E.B. Sandell. 1952. *Textbook of Quantitative Inorganic Analysis.* Macmillan Co., New York, NY.

Kononova, M.M. 1961. *Soil Organic Matter.* T.Z. Nowakowski and G.A. Greenwood (transl.). Pergamon Press, Oxford. England

Krauskopf, K.B. 1973. Geochemistry of micronutrients. p.7-40. In: *Micronutrients in Agriculture,* J.J. Mortvedt (ed). Soil Sci. Soc. Am., Madison, WI.

Krumbein, W.C., and F.J. Pettijohn. 1938. *Manual of Sedimentary Petrology.* Appleton-Century Co., Inc., New York, NY.

Kubota, J., and E.E. Cary. 1982. Cobalt, molybdenum, and selenium. p.485-500. In: *Methods of Soil Analysis,* Part 2, A.L. Page, R.H. Miller, and D.R. Keeney (eds). 2nd ed. Agronomy series No.9. Am. Soc. Agronomy, Inc., Publ., Madison, WI.

Lakatos, B., J. Meisel, and G. Mady. 1977. Biopolymer metal

complex systems. I. Experiments for the preparation of high purity peat humic substances and their metal complexes. Acta Agron. Acad.Sci. Hung. 26:259-271.

Lindsay, W.L. 1973. Inorganic phase equilibria of micronutrients in soils. p.41-57. In: *Micronutrient in Agriculture*, J.J. Mortvedt, P.M. Giordano, and W.L. Lindsay (eds). Soil Sci. Soc. Am., Inc., Madison, WI.

Lindsay, W.L., and W.A. Norvell. 1978. Development of DTPA soil test for zinc, iron, manganese, and copper. Soil Sci. Soc. Am. J. 42:421-428.

Lobartini, J.C., K.H. Tan, L.E. Asmussen, R.A. Leonard, D. Himmelsbach, and A.R. Gingle. 1989. Humic matter isolated from soils and water by the XAD-8 resin and conventional NaOH methods. Comm. Soil Sci. Plant Anal. 20:1453-1477.

Mackenzie, R.C. 1969. Nomenclature in thermal analysis. Talanta 16: 1227-1230.

Mackenzie, R.C. (ed). 1970. *Differential Thermal Analysis*. Academic Press, London and New York, NY.

Martin, R.T. 1955. Reference chlorite characterization for chlorite identification in soil clays. Clays Clay Min. 3:117-145.

Matejka, J. 1922. Thermal analysis as a means of detecting kaolinite in soils. Chem. Listy 16:8-14.

McGeorge, W.T. 1944. Base-exchange - pH relationships in semiarid soils. Soil Sci. 59:271-275.

McKeague, J.A., and M.G. Cline. 1963. Silica in soil solution. II. The adsorption of monosilicic acid by soil and other substances. Can. J. Soil Sci. 43:83-96.

McLean, E.O., and M.E. Watson. 1985. Soil measurements of plant-available potassium. p.277-308. In: *Potassium in Agriculture*, R.D. Munson (ed). Am. Soc. Agronomy, Madison, WI.

Mehlich, A. 1960. Charge characterization of soils. Trans. Int. Soil Sci. Conf., Madison, WI, Vol II/III: 292-302.

Mehlich, A. 1981. Charge properties in relation to sorption and desorption of selected cations and anions. In: *Chemistry in the Soil Environment*, D.E. Baker (org.comm.),. ASA Spec. Publ. No.40, Am. Soc. Agronomy, Madison, WI.

Mehlich, A. 1984. Mehlich 3 soil test extractant: A modification of Mehlich 2 extractant. Comm. Soil Sci. Plant Anal. 15:1409-1416.

Mehta, N.C., J.O. Legg, C.A.I. Goring, and C.A. Black. 1954. Determination of organic phosphorus in soils: I. Extraction methods. Soil Sci. Soc. Am. Proc. 18:443-449.

Miller, F.A., and C.H. Wilkens. 1952. Infrared spectra and characteristic frequencies of inorganic ions. Anal. Chem. 24: 1253-1294.

Mohr, E.C.J., and F.A. Van Baren. 1960. *Tropical Soils*. Les editions A. Manteau S.A., Brusells.

Mortenson, J.L., D.M. Anderson, and J.L. White. 1965. Infrared spectrometry. p.743-770. In: *Methods of Soil Analysis*, Part 1, C.A. Black (ed-in-chief). Agronomy series No.9. Am. Soc. Agronomy, Inc., Publ., Madison, WI.

Munson, R.D. (ed). 1985. *Potassium in Agriculture*. Am. Soc. Agronomy, Madison, WI.

Murphy, C.B. 1962. Differential thermal analysis. A review of fundamental developments in analysis. Anal. Chem., Review Issue 34: 298R - 301R.

Nelson, D.W., and L.E. Sommers. 1982. Total carbon, organic carbon, and organic matter. p.539-579. In: *Methods of Soil Analysis*, Part 2, A.L. Page, R.H. Miller, and D.R. Keeney (eds),. Agronomy series No.9. Am. Soc. Agronomy and Soil Sci. Soc. Am., Inc., Publ., Madison, WI.

Nommik, H. 1965. Ammonium fixation and other reactions involving a nonenzymatic immobilization of mineral nitrogen in soil. p.198-258. In: *Soil Nitrogen*, W.V. Bartholomew and F.E. Clark (eds). Agronomy series No. 10. Am. Soc. Agronomy, Inc., Publ., Madison, WI.

Norvell, W.A. 1973. Equilibria of metal chelates in soil solution. p.115-138. In: *Micronutrient in Agriculture*, J.J. Mortvedt, P.M. Giordano, and W.L. Lindsay (eds). Soil Sci. Soc. Am.,

References

Inc., Madison, WI.

Olsen, S.R., and L.E. Sommers. 1982. Phosphorus. p.403-430. In: *Methods of Soil Analysis*, Part 2, A.L. Page, R.H. Miller, and D.R. Keeney (eds). 2nd ed. Agronomy series No.9, Am. Soc. Agronomy and Soil Sci. Soc. Am., Inc., Madison, WI.

Orioli, G.A., and N.R. Curvetto. 1980. Evaluation of extractants for soil humic substances. 1. Isotachophoretic studies. Plant and Soil 55:353-361.

Pauling, L. 1964. *College Chemistry*. Freeman and Co., San Francisco, CA.

Peech, M. 1945. Determination of exchangeable cations and exchange capacity of soils - Rapid micromethods utilizing centrifuge and spectrophotometer. Soil Sci. 59:28-38.

Peech, M., L.T. Alexander, L.A. Dean, and J.F. Reed. 1947. Methods of soil analysis for soil-fertility investigations. U.S. Dept. Agric. Circ. No.787.

Peterson, R.G., and L.D. Calvin. 1986. Sampling. p.33-51. In: *Methods of Soil Analysis*, Part 1, A. Klute (ed). Agronomy series No.9. Am. Soc. Agronomy and Soil Sci. Soc. Am., Inc., Publ., Madison, WI.

Pregl, F. 1949. *Quantitatief Organische Mikro Analyse*. Neubearbeitet von Dr. H. Roth, Wien. Translated by J. Grant, 4th English edition. P. Blakiston's Son & Co., Philadelphia.

Puri, A.N., and A.G. Asghar. 1937. Influence of salt and $CaCl_2$-water ratio on pH value of soils. Soil Sci. 46:249-257.

Reed, J.F., and J.A. Rigney. 1947. Soil sampling from fields of uniform and non-uniform appearance and soil types. J. Am. Soc. Agronomy 39: 26-40.

Rhoades, J.D. 1982. Cation exchange capacity. p.149-157. In: *Methods of Soil Analysis*, Part 2, A.L. Page, R.H. Miller, and D.R. Keeney (eds). Agronomy series No.9. Am.Soc. Agronomy and Soil Sci. Soc. Am., Inc., Publ., Madison,WI.

Robinson, W.O. 1930. Method and procedure of soil analysis used in the Division of Soil Chemistry and Physics. U.S. Dept. Agric. Cir. 139 (revised 1939).

Russell, M.B., and J.I. Haddock. 1940. The identification of clay minerals in five Iowa soils by the thermal method. Soil Sci. Soc. Am. Proc. 5:90-94.

Sayre, J.D. 1952. Accumulation of radio-isotopes in corn leaves. Ohio Agri. Expt. Sta. Res. Bull. 723

Schnitzer, M. 1974. The methylation of humic substances. Soil Sci. 117:94-102.

Schnitzer, M. 1982. Organic matter characterization. p.581-594. In: *Methods of Soil Analysis*, Part 2, A.L. Page, R.H. Miller, and D. R. Keeney (eds). Agronomy series No.9. Am. Soc. Agronomy and Soil Sci. Soc. Am., Inc., Publ., Madison, WI.

Schnitzer, M., and S.U. Khan. 1972. *Humic Substances in the Environment*. Marcel Dekker, New York, NY.

Schnitzer, M., and H. Kodama. 1972. Differential thermal analysis of metal-fulvic acid salts and complexes. Geoderma 7:93-103.

Schnitzer, M., D.A. Shearer, and J.R. Wright. 1959. A study in the infrared of high-molecular weight organic matter extracted by various reagents from a Podzolic B horizon. Soil Sci. 87: 252-257.

Schnitzer, M., and S.I.M. Skinner. 1965. Organo-metallic interactions in soils: 4. Carboxyl and hydroxyl groups in organic matter and metal retention. Soil Sci. 99: 279-284.

Schofield, R.K., and A.W. Taylor. 1955. The measurement of soil pH. Soil Sci. Soc. Am. Proc. 19:164-167.

Schollenberger, C.J. 1927. A rapid approximate method for determining soil organic matter. Soil Sci. 24:65-68.

Schollenberger, C.J. 1945. Determination of soil organic matter. Soil Sci. 59:53-56.

Schollenberger, C.J., and F.R . Dreibelbis. 1933. Analytical methods in base exchange investigations of soils. Soil Sci. 30:161-173.

Schollenberger, C.J., and R.H. Simon. 1945. Determination of exchange capacity and exchangeable bases in soil - Ammonium acetate method. Soil Sci. 59: 13-24.

Schultze, D. 1969. *Differential Thermo-analyses*. Verlag Chemie GmbH. Weinheim/Bergstr.

Sherman, G.D., and P.M. Harmer. 1943. The effect of manganese sulfate on several crops growing on organic soil when applied as a stream or spray on the crop. Soil Sci. Soc. Am. Proc. 8:334-340.

Skoog, F. 1940. Relations between zinc and auxin in the growth of higher plants. Am. J. Bot. 27:939-951.

Smith, A.L. 1979. *Applied Infrared Spectroscopy: Fundamentals, Techniques and Analytical Problem-Solving*. Chemical Analysis, Vol.54. John Wiley & Sons, New York, NY.

Smith, K.A., and A. Scott. 1990. Continuous-flow, flow-injection, and discrete analysis. p.183-227. In: *Soil Analysis, Modern Instrumental Techniques*, K.A. Smith (ed). 2nd ed. Marcel Dekker, Inc., New York, NY.

Snedecor, G.W., and W.G. Cochran. 1974. *Statistical Methods*. Iowa Univ. Press, Ames, Iowa.

Soil Survey Staff. 1951. *Soil Survey Manual*. U.S.D.A. Handbook No.18. U.S. Govern. Printing Office, Washington, D.C.

Soil Survey Staff. 1972. *Soil Survey Laboratory Methods and Procedures for Collecting Soil Samples*. Soil Survey Investigations Report No.1. U.S.D.A., S.C.S. U.S. Govern. Printing Office, Washington, D.C.

Soil Survey Staff. 1975. *Soil Taxonomy. A Basic System of Soil Classification for Making and Interpreting Soil Surveys*. Agric. Handbook No. 436. U.S. Govern. Printing Office, Washington, D.C.

Soltanpour, P.N., R.L. Fox, and R.C. Jones. 1988. A quick method to extract organic phosphorous from soils. Soil Sci. Soc. Am. J. 51:255-256.

Spangler, M.G., and R.L. Handy. 1984. *Soil Engineering*. Harper and Row, Publ., New York, NY.

Stevenson, F.J. 1965[a]. Gross chemical fractionation of organic matter. p.1409-1421. In: *Methods of Soil Analysis*, Part 2, C.A.Black (ed-in-chief). Agronomy series No. 9. Am. Soc.

Agronomy, Inc., Madison, WI.
Stevenson, F.J. 1965[b]. Origin and distribution of nitrogen in soil. p.1-42. In: *Soil Nitrogen*, W.V. Bartholomew and F.E. Clark (eds). Agronomy series No. 10. Am. Soc. Agronomy, Inc., Publ., Madison, WI.
Stevenson, F.J. 1979. Humus. p.195-205. In: *The Encyclopedia of Soil Science*, Part 1, R.W. Fairbridge and C.W. Finkle, Jr. (eds),. Dowden, Hutchinson and Ross, Inc., Stroudsburg, PA.
Stevenson, F.J. 1982. *Humus Chemistry, Genesis, Composition, Reactions*. John Wiley and Sons, Inc., New York, NY.
Stevenson, F.J. 1986. *Cycles of Soil, Carbon, Nitrogen, Phosphorus, Suffur, Micronutrients*. John Wiley and Sons, Inc., New York, NY.
Stevenson, F.J. 1994. *Humus Chemistry. Genesis, Composition, Reactions*. 2nd ed. John Wiley & Sons, New York, NY.
Stewart, J.E. 1970. *Infrared Spectroscopy: Experimental Methods and Techniques*. Marcel Dekker, Inc., New York, NY.
Swinehart, J.S., and R.C. Gore. 1968. Forensics. Perkin-Elmer Infrared Applications Study No.6. Perkin-Elmer Corp., Norwalk, Connecticut.
Tabatabai, M. 1982. Sulfur. p.501-538. In: *Methods of Soil Analysis*, Part 2, A.L. Page, R.H. Miller, and D.R. Keeney (eds). 2nd ed. Agronomy series No.9. Am. Soc. Agronomy and Soil Sci. Soc. Am., Inc., Publ., Madison, WI.
Tan, K.H. 1964. The Andosols in Indonesia. Meeting on the Classification and Correlation of Soils from Volcanic Ash, Tokyo, Japan. FAO World Soil Resources Report No.14, p.30-35.
Tan, K.H. 1975. Infrared absorption similarities between hymatomelanic acid and methylated humic acid. Soil Sci. Soc. Am. Proc. 39:70-73.
Tan, K.H. 1978. Formation of metal-humic acid complexes by titration and their characterization by differential thermal analysis and infrared spectroscopy. Soil Biol. Biochem. 10:

123-129.

Tan, K.H. 1982. *Principles of Soil Chemistry*. Marcel Dekker, Inc. New York, NY.

Tan, K.H. 1990. *Basic Soils Laboratory*. Alpha edition. Burgess Publ.Co., Minneapolis, Minnesota.

Tan, K.H. 1992. Humic acids. In: *The Encyclopedia of Soil Science*, Part 1, C.W. Finkl, Jr.(ed), Second edition. Van Nostrand Reinhold, New York, NY.

Tan, K.H. 1994. *Environmental Soil Science*. Marcel Dekker, Inc., New York, NY.

Tan, K.H., E.R. Beaty, R.A. McCreery, and J.B. Jones. 1975. Differential effect of Bermuda and Bahiagrasses on soil chemical characteristics. Agronomy J. 67:407-411.

Tan, K.H., and A. Binger. 1986. Effect of humic acid on aluminum toxicity in corn plants. Soil Sci. 14:20-25.

Tan, K.H., and P.S. Dowling. 1984. Effect of organic matter on CEC due to permanent and variable charges in selected temperate region soils. Geoderma 32:89-101.

Tan, K.H., B.F. Hajek, and I. Barsjad. 1986. Thermal analysis techniques. p.151-183. In: *Methods of Soil Analysis*, Part 1, A. Klute (ed). Agronomy series No.9. Am. Soc. Agronomy and Soil Sci. Soc. Am., Inc., Publ., Madison, WI.

Tan, K.H., and D.L. Henninger. 1993. Mineralogy, chemical composition, and interaction of lunar simulants with humic acids. Comm. Soil Sci. Plant Anal. 24:2479-2492.

Tan, K.H., R.A. Leonard, L.E. Asmussen, J.C. Lobartini, and A.R. Gingle. 1990. The geochemistry of black water in selected coastal streams of the southeastern United States. Comm. Soil Sci. Plant Anal. 21:1999-2016.

Tan, K.H., J.C. Lobartini, D.S. Himmelsbach and L.E. Asmussen. 1991. Composition of humic acids extracted under air and nitrogen atmosphere. Comm. Soil Sci. Plant Anal. 22:861-877.

Tan, K.H. and R.A. McCreery. 1970. The infrared identification of a humo-polysaccharide ester in soil humic acid. Comm. Soil Sci. Plant Anal. 1(2):75-84.

Tan, K.H. and P.S. Troth. 1981. Increasing sensitivity of organic matter and nitrogen analysis using soil separates. Soil Sci. Soc. Am. J. 45:574-577.

Tisdale, S.L., and W.L. Nelson. 1993. *Soil Fertility and Fertilizers*. Macmillan Co., New York, NY.

Tisdall, A.L. 1951. Comparison of methods of determining apparent density of soils. Austr. J. Agric.Res. 2:349-354.

Thomas, G.W. 1982. Exchangeable cations. p.159-165. In: *Methods of Soil Analysis*, Part 2, A.L. Page, R.H. Miller, and D.R. Keeney (eds). Agronomy series No. 9. Am. Soc. Agronomy and Soil Sci. Soc. Am., Inc., Publ., Madison, WI.

Thomson, A.P., D.M.L. Duthie, and M.J. Wilson. 1972. Randomly oriented powders for quantitative identification by x-ray diffraction. Clay Mineral. 9: 345-348.

Thurman, E.M., and R.L. Malcolm. 1981. Preparative isolation of aquatic humic substances. Environ.Sci.Techn. 15:463-466.

Uchiyama, N., and N. Onikura. 1955. Investigations on clay formation in paddy soils. I. Variation in silica and alumina soluble in dilute sodium carbonate solutions during one year. J. Soil Sci. (Tokyo) 25:291-298.

Ulrich, A. 1948. Plant analysis methods and interpretation of results. p.157-198. In: *Diagnostic Techniques for Soils and Crops*, H.B. Kitchen (ed). Am. Potash Inst., Washington, D.C.

Van der Marel, H.W., and H. Beutelspacher. 1976. *Atlas of Infrared Spectroscopy of Clay Minerals and their Admixtures*. Elsevier Sci. Publ., Amsterdam.

Van Schuylenborgh, J. 1954. The effect of air-drying of soil samples upon some physical soil properties. Neth. J. Agri. Sci. 2:50-57.

Walkley, A. 1947. A critical examination of a rapid method for determining organic carbon in soils. Effect of variations in digestion conditions and inorganic soil constituents. Soil Sci. 63:251-263.

Wall, L.L., C.W. Gehrke, T.E. Neuner, R.D. Cathey, and P.R. Rexroad. 1975. Total protein nitrogen evaluation and compar-

References

ison of four different methods. J. Assoc. Off. Anal. Chem. 58:811-817.

Whetsel, K. 1968. Infrared spectroscopy. Chem. Engin. News 46:82-96.

White, J.L., and C.B. Roth. 1986. Infrared spectrometry. p.291-330. In: *Methods of Soil Analysis*, Part 1, A. Klute (ed). Agronomy series No. 9. Am. Soc. Agronomy and Soil Sci. Soc. Am., Inc., Publ., Madison, WI.

Whittig, L.D., and W.R. Allardice. 1986. X-ray diffraction techniques. p.331-362. In: *Methods of Soil Analysis*, Part 1, A. Klute (ed). Agronomy series No. 9. Am. Soc. Agronomy and Soil Sci. Soc. Am., Inc., Publ., Madison, WI.

Willard, H.H., and H. Diehl. 1943. *Advanced Qualitative Analysis*. Van Nostrand, New York, NY.

Willard, H.H., L.L. Merritt, Jr., and J.A. Dean. 1974. *Instrumental Methods of Analysis*. Van Nostrand, New York, NY.

Wirjodihardjo, M.W., and K.H. Tan. 1964. *Ilmu Tanah* (Soil Science, in Indonesian). Part II. Pradnjaparamita II, Jakarta.

INDEX

Absorbance, 246
  relationship with concentration, 248
Accuracy, 27, 29
  checking for, 35
  definition of, 27
Acid sulphate soil, 184
Acidity,
  active, 98
  exchange, 98
  potential, 98
  reserve, 98
  total, 98
Acre-furrow-slice,
  definition of, 87
  relationship with bulk density, 87
Active acidity, 98
Adenosine diphosphate, ADP, 154
Adenosine triphosphate, ATP, 154, 183
Adhesion, 56
Air drying soils, 17
  effect of, 18-20

Albite, 175, 200
Allophane,
  DTA identification of, 352, 354
Aluminum,
  available, 191
  colorimetric determination of, 193
  content in soils, 189, 190
  determination of, 191
  exchangeable, 134, 190
  extraction of exchangeable, 194
  extraction of total, 192
  hexahydronium, 190
  importance in soil taxonomy, 190
  oxides, 189
  sources of, 189
Ammonium, NH_4^+,
  colorimetric determination of, 145, 147
  determination by Kjeldahl, 144

Index

determination of exchangeable, 138-139
exchangeable, 138, 148
extraction of exchangeable, 149
fixation, 137
fixed, 107, 137
Analytical error, 1
Ascorbic acid sulfomolybdo phosphate blue color method, 161
Atomic absorption, 255, 256
  atomizer and flame in, 261, 264
  chemical interference in, 266-267
  detection limits in, 276-277
  gas sample in, 266
  interference by accompanying elements, 267
  interference by acids, 268
  ionization interference in, 268
  liquid sample in, 265
  main components of, 258-260
  physical interference in, 269
  polarized light in, 261
  primary light source in, 261
  principles of, 257
  solid sample in, 265
  solvent interference, 267
  temperature in, 263-265
Atomic emission, 269
  principles of, 269
Atomic fluorescence spectroscopy, 255, 256
Attapulgite,
  infrared spectrum of, 296, 297
Available water,
  definition of, 58
  determination of, 62
  pF value of, 58
Axenite, 212
Azurite, 208

Barium hydroxide method,
  in determination of total acidity, 238-239
  reagents in, 238
  procedure, 238
Bauxite, 189
Beer's law, 244, 246
Bias,
  additive, 31
  checking for, 31
  definition of, 30
  measurement of, 31
  multiplicative, 31-32
  sampling, 31
  scientific, 31
Biotite,
  infrared spectrum of,

296, 297
Blue silicomolybdous acid
  method, 203
Bolometers, 284
Boltzmann equation,
  in atomic emission, 270
Borax, 211
Boric acid, H_3BO_3, 211
Boron, 211
  available, 215
  content in soils, 212
  deficiency of, 213
  determination of, 213
  effect on plants, 212
  hot water extraction of, 215
  ionic form of, 212
  silicate minerals, 211
  sodium carbonate fusion of, 214
  solubility of, 212
  source of, 211
  toxicity, 213
Bragg's law, 251, 252
  in XRD, 306
Braunite, Mn_2O_3, 204
Brij-35, 145
Buffer capacity, 98
Bulk density,
  definition of, 86
  determination of, 86, 88-89, 90-91
  types of, 87
  units of, 87
  values of, 89

Calcite, 178
Calcium,
  bicarbonate, 179
  content in soils, 179
  determination of, 179-180
  effect on soil structure, 178
  source of, 178
Calcium acetate method,
  in determination of COOH, 240-241
  in determination of phenolic-OH, 241
  procedure, 241
  reagents in, 240
Capillary force, 56
Carbonation, 179
Catalyst,
  in Kjeldahl, 142, 143
Cat clays, 184
Cations,
  adsorbed, 114
  exchangeable, 114
  influence on soils, 116, 117
Cation exchange capacity,
  definition of, 114-115, 122
  determination of, 123, 126-128, 130
  effective, ECEC, 116
  effect of hydrolysis of displacing ions on, 121
  effect of K-fixation on,

Index

121
effect of precipitation reaction on, 121
factors affecting, 117-123
produced by permanent charges, CEC_p, 116
produced by variable charges, CEC_v, 116
of inorganic colloids, 119
of organic colloids, 119
reaction, 114-115
total, CEC_t, 116
units of, 115
Chemical analysis,
definition of, 53
types of, 53
Clay,
definition of, 73-74
separation of, 83-84
Clod method,
calculations in, 91
in bulk density determination, 88, 90
cm of water,
conversion into bars, 57
conversion into pF, 57
definition of, 57
values for soil moisture constants, 58
Coesite, 199
Cohesion, 56
Colemanite, 212
Colorimetry,
definition of, 242
Compositing,

definition of, 8
objectives of, 8-9
Compton scatter, 301
Constructive reinforcement, 251
Copper, 207
auto-reduction-oxidation of, 208
cadmium reductor, 149
content in soils, 208
determination of, 209-210
effect on plants, 208
extraction of available, 211
extraction of exchangeable, 210
source of, 207
toxicity, 209
Coupling agent in determination of NO_2, 152
Crystal lattice, 305
Crystobalite, 199

Deferration, 308, 310
Destructive interference, 251
Detector,
noise, 254
photoelectric, 254
Deuterium lamp, 250
Devarda alloy, 149, 152
Deviation,
definition of, 29
standard, 30, 38
Diazotizing solution, 152

Differential dissolution in DTA, 354
Differential scanning calorimetry, DSC, 329
  modulated, 329
Differential thermal analysis, DTA, 329
  cation saturation in, 340
  enthalpy changes in, 332
  furnace, 332
  hydration and solvation in, 344-345
  of clay, 338-339
  of sand, 337
  of silt, 338
  of whole soil, 335-336
  Ohm's law in, 333
  peak temperatures in, 331
  reactions in, 330-331
  specimen holder, 332, 334
  thermal break in, 331
  thermocouples in, 330, 333
  well holder in, 333
Dispersion,
  in soil texture analysis, 76
  in spectrophotometry, 252, 253
Dithionite-citrate-bicarbonate method, DCB, 199, 310
Dolomite, 178, 181
  dissolution of, 181
Double acid procedure, 158
Dry combustion method of organic matter, 226-227
d-Spacings in XRD,
  as affected by glycolation, 309
  as affected by heat, 309
  as affected K-saturation, 309
  as affected by Mg-saturation, 309
  conversion table, 383
  of soil minerals, 323-326
Dumas method, 138, 226

EDAX, 50
Edaphology, 42
  soil analysis in, 43-47
Electrical resistance block, see gypsum block, 61
Electrode,
  calomel, 103, 105
  combination, 104-105
  glass, 103, 105
  indicator, 103
  in pH measurement, 103-104
  liquid junction potential of, 108
  reference, 103, 105
Emerald,
  Brazilian, 212

Endothermic peak,
  effect of heating rate on, 350
  effect of sample size on, 346
  height versus gibbsite content, 360
  in DTA, 330, 331
  of clays in DTA, 355-357
  of humic acid in DTA, 340-341
  temperatures of clays, 355-357
Endpoint of titration, 53
Equivalent point, 53
Equivalent weight,
  in acid-base reaction, 54
  in complex reaction, 54
  in redox reaction, 54-55
Error,
  absolute, 28
  analytical, 34
  definition of, 27-28
  determinate, 32
  indeterminate, 32, 34
  instrumental, 32
  operative, 32-33
  personal, 33
  relative, 28
  standard, 30, 38
  titration, 53
Exchangeable Al^{3+}
  determination of, 134
Exchangeable H^+, 118,
  determination of, 132, 134
Exchange acidity,
  see Potential acidity
Excitation,
  by electrical discharge, 272-273
  by radiation, 273
  flame, 272
  laser microprobe, 273-274
Exothermic peak,
  in DTA, 330, 331
  of clays in DTA, 355-357
  temperatures of clays, 355-357
Extinction coefficient, 247
Extracting solutions,
  ammonium acetate, NH_4OAc, 45
  Bray P-1, 45, 157, 166
  Mehlich's 1, 45, 157, 167
  Mehlich's 2, 45, 157
  Mehlich's 3, 45, 157
  Morgan's, 45, 157
  Olsen's, 157, 166
  universal, 45

Feldspar,
  potash, 171
  sodic, 175
Ferrimolybdate,
  $Fe_2(MoO_4)_3 \cdot 8H_2O$, 216
Field capacity, 58
  measurement of, 67-68

of available water, 62
First order diffraction, 307
Fixed NH_3, 137
Flame emission spectroscopy, 255, 271
Flame photometry, 272
Fluoro-boric acid digestion, 159-160
  in determination of,
    Al, 191
    Cu, 210
    Fe, 196
    K, 173
    Mn, 206
    Mo, 218
    Na, 177
    Si, 203
    Zn, 220, 221
Fluorescence, 256
Fool's gold, 194
Fulvic acid,
  DTA identification of, 341, 357
  infrared spectrum, 287, 288, 289
  purification of, 236

Gibbsite, 189
  DTA identification of, 353
  infrared spectrum of, 296, 297-298
  XRD identification of, 353

Globar glower, 284
Glycolation in XRD, 310
Goethite,
  DTA identification of, 353-354
  XRD identification of, 316-317, 322
Gravimetric method, 53
Grinding,
  equipment, 21-22
  errors in, 23-24
  loss or gain in moisture during, 24-25
  of samples, 21
  oxidation in, 24
  purpose of, 21-22
Griess-Ilosvay method, 149-150, 152-153
Gross sample, 1, 2, 4
Gypsum, 178, 183
  block method, 61

Halloysite,
  DTA identification of, 351, 352
Heated graphite atomizer, HGA, 256
Heat treatment in XRD, 311
Hematite, 189, 194
Hemimorphite, $Zn_4(Si_2O_7)(OH)_2 \cdot H_2O$, 219
Hollow cathode lamp, 258-

Index

259
  combination, 260
  multielement, 260
  single element, 260
Humic acid,
  DTA identification of, 341, 357
  DTA identification of metal complexes of, 341, 358
  infrared band assignment of, 291
  methylation of, 291-292
  purification of, 235
Humic matter, 232
  carboxyl content of, 239-240
  determination of, 232-233
  extraction of, 234-235
  extraction of aquatic, 236
  extraction of HA and FA, 234-235
  phenolic-OH content of, 239-240
Humin infrared spectrum, 289, 290
Hydrodynamic radius,
  of cations, 120
    effect on cation exchange, 120
Hydrogen lamp, 250
Hydrogen sulphite, H_2S, 183
  oxidation of, 184
Hydrologic cycle, 56
Hydrolyzable unknown form of N, HUN, 136
Hydrometer, 76, 77
Hydroxy-Al interlayered vermiculite,
  XRD identification of, 315, 320
Hydrozincite, $Zn_5(CO_3)_2(OH)_6$, 219
Hygroscopic coefficient, 58
Hymatomelanic acid,
  infrared spectrum of, 289, 290
Hypochlorite method, 145

Ignition method,
  extraction of sample in, 165
  in P determination, 164
Illite,
  infrared spectrum of, 295, 297
  XRD identification of, 315, 318
Ilmenite, 194
Independent analysis, 37
Indicator,
  constant, 101
  definition of, 99
  duplex, 101
  reactions, 100
  types of, 102
  universal, 101

use in pH measurement, 99-100
Individuals in sampling, 4
Indophenol blue color method, 145
Inductively coupled plasma, ICP, spectroscopy, 272, 274
  detection limits, 276-277
  torch, 275
Infrared active, 280
Infrared gas cell, 285
Infrared inactive, 280
Infrared radiation,
  types of, 279
Infrared spectroscopy,
  absorption bands of functional groups in clays, 294
  application of, 278
  clay film technique in, 292
  dispersive, 284
  finger print region in, 287
  Fourier transform, 284
  group frequency region in, 287
  Hooke's law in, 280-281
  in humic matter, 288
  KBr pellet technique in, 286, 292
  Lambert and Beer's law in, 283
  molecular vibrations in, 280-282
  mull technique in, 292
  samples in, 285
  types of dies in, 286
  units in, 279
Inositol phosphate, 156
Internal consistency, 39
  method of, 40
Ionic balance, 36
Ionization potential, 270-271
Iron, 194
  availability index of, 197
  content in soils, 195
  determination of, 195-196
  DTPA extractable, 197
  effect of airdrying on Fe content, 20
  extraction of available, 198
  extraction of free and amorphous, 199
  sources of, 194
Irtran window, 285

Kaolinite,
  DTA of air dry, 345
  DTA of dried, 345
  DTA thermogram of, 342, 351, 352
  infrared spectrum of, 294-295
  XRD identification of,

Index

315, 316
Kelmate, 143, 146
Kernite, 212
Kjeldahl method, 138-139
  block digester in, 140
  digestion phase in, 141
  digestion time in, 144
  distillation phase in, 139
  reagents in, 141-142
  sample preparation in, 140
  sample size in, 140
Kjeltabs, 143, 146

Lambert and Beer's law, 246-247
  in atomic absorption, 257
Lambert and Bouger's law, 244, 245-246
Lattice spacing, 314
Leaf testing, 45
Liquid junction potential, see electrode
Lignin infrared spectrum, 289, 290
Lime potential,
  definition of, 112
  determination of, 112-113
Lime requirement, 109
  determination of, 109-110
Limonite, 194

Magnesium,
  content in soils, 181
  determination of, 181-183
Magnetite, 194
Malachite, 208
Manganese,
  content in soils, 204
  critical concentration, 205
  determination, 205
  effect of air drying on content of, 204
  effect on plants, 205
  exchangeable, 204
  extraction of available, 207
  extraction of exchangeable, 206-207
  source of, 181
  toxicity, 205
Manganite, MnO(OH), 204
Maximum retentive capacity, 58
Max Planck's constant, 305
Mechanical analysis, 76
Mechanics of sampling, 9-10
Mg-saturation in XRD, 310
Micromorphology, 49
Micromorphometry, 50
Micropedology, 49
Moisture equivalent, 58

Molecular vibrations,
  bending, 280-281, 287
  of aromatic compounds, 282
  of polyatomic systems, 287
  stretching, 280-281, 287
Molybdenite, MoS_2, 216
Molybdenosis, 217
Molybdenum,
  available, 217
  content of, 216
  deficiency, 216
  ionic form, 216
  source of, 216
  toxicity, 216
Monochromator, 242, 251
  grating, 251, 252
  prism, 251, 252
Mull, 285
Muscovite infrared spectrum, 295, 297

Near infrared reflectance spectroscopy, 138
Neat compound, 285
Nephelometry, 244
Nernst glower, 284
Nitrate, NO_3^-, 137
  direct determination of, 150
  extraction of, 150
  indirect determination of, 149

Nitrite, NO_2^-, 152
  direct determination of, 152
  extraction of, 152
  indirect determination of, 153
Nitrogen,
  content in soils, 135
  determination of total, 138
  effect of drying on content of, 19
  inorganic, 137
  major forms of, 136
  organic, 136
  total, 138
Nitrogen cycle, 137
Nucleic acid, 156
Nujol, 285

Opal,
  infrared spectrum of, 296, 298
Organic carbon, C_{org},
  determination, 226, 229
  effect of air drying on, 18

Parallel determination, 39
  independent, 39
Particle density, 87
  determination of, 91-92, 93
  values of, 92

Index

Particle size distribution,
  determination of, 77, 78-80, 81-84, 84
Pedon, 2
Pedology, 42,
  soil analysis in, 48-51
Personal judgement error, 33
pF, 57
  values of, 58
pH, see soil pH
Phenate method,
  see Indophenol blue color method
Phenol disulfonic acid method, 150-157
Phospholipid, 156
Phosphorus,
  anion resin extractable, 159, 168-169
  available, 154, 157
  content in soil, 154-155
  determination
    by anion resin, 168-169
    by Bray's method, 157, 166
    by isotopic dissolution, 169
    by Mehlich's method, 167-168
    by Olson's method, 168-169
  importance in plants, 153
  effect of airdrying on
    content of, 19
  inorganic, 154-156
  organic, 154-157
  soluble in water, 158
Photodetector, 284
Phototubes, 254
Photovoltaic cell, 254
Physicochemical analysis, 51-52
Phytic acid, 156
Plagioclase, 175
Plant testing, 45, 47
Polypedons, 2
Pore space,
  definition of, 94
  determination of, 95
  types of, 94
Potash,
  see Potassium
Potassium,
  availability indices, 171
  available, 172
  effect of airdrying on content of, 20
  extraction of exchangeable, 174
  fixed, 172
  nonexchangeable, 172
  total, 173
  sources of, 170-171
Potassium dichromate in C_{org} determination, 229-230
Potassium permanganate in C_{org} determination,

230
Potassium saturation in XRD, 310
Phenate method, 145
Phenol disulphonic acid method, 150-151
Phenol-Na-hypochlorite method, 145
Potassium sulfate, 141, 143
Potential acidity, 98
Powellite, $CaMoO_4$, 216
Precision,
  checking for, 38
  definition of, 27
  measure of, 28-29
Prejudice, 33
Pressure plate, 62, 64
Pyrite, 183, 194
Pyrolusite, MnO_2, 204

Qualitative analysis, 51
Quantitative analysis, 51-52
Quantitative XRD, 327
  internal standard method in, 327
  matrix flushing in, 327
  reference intensity ratio in, 328
Quartering of sample, 25
Quartz, 199-200
  DTA analysis of, 335, 337

Rapid field test, 47
Rayleigh scatter, 301
Representative sample, 1
Resolution in spectrophotometry, 252
Roentgen radiation, 299

Salicylate-hypochlorite method, 145
Salicylic acid, 141
Sample,
  in soil testing, 44
  preparation of, 21
  preservation of, 17
  representative, 1-2
  required number of, 14-16
  splitting, 26
Sampling, 9-10
  accuracy of, 1, 12-13
  biosequence in, 11
  chronosequence in, 11
  climosequence in, 11
  deep, 10
  disturbed soil, 10
  error, 1
  lithosequence in, 11
  methods of, 1, 5-9
  rule of homogeneity in, 4-5, 7, 8
  rule of proportionality in, 7
  shallow, 9-10
  simple random, 6

Index

size, 12
stratified, 7
soil profile, 10
systematic, 7
time, 11-12
toposequence in, 11
unit, 4
volume, 4
Sand,
  definition of, 73
  separation of, 82
Saran resin, 90
Sedimentation, 76
Sequential extraction of organic P, 162
Sieving,
  contamination in, 24
  errors in, 23-24
  of sample, 22
  purpose of, 22
  sieve requirement in, 22-23
Significant figures in
  weighing, 40-41
  calculations, 41
Silica, 199
  gel, 201
Silicic acid,
  mono, 201
  ortho, 201
Silicon,
  content in rocks and soils, 200-201
  determination of, 201
  sources of, 199

Silt,
  definition of, 73
  separation of, 83
Siloxane bond, 200
Simple harmonic motion, 280
Smectite,
  DTA identification of, 51-352
  DTA of cation-saturated, 343
  infrared spectrum of, 295, 297
  XRD of, 315, 317
Smithsonite, $ZnCO_3$, 219
Sodium,
  content in soils, 176
  determination of, 177
  exchangeable, 176
  exchangeable percentage, ESP, 176
  extraction of exchangeable, 178
  source of, 175
Sodium nitroprusside, 145-147
Sodium salicylate-hypochlorite method, 147
Sodium sulfate, 141
Soil density,
  definition, 86
  determination of, 86
  importance of, 86
  relation to specific gravity, 87-88

types of, 86-87
Soil formation factors, 2
Soil horizon, 4
Soil moisture content, 56
  determination of available, 58, 62-66
  determination of total, 58, 59-60
  effect of airdrying on, 18
Soil moisture constant, 58
Soil moisture tension, 57
  determination of, 68, 69-71
  effect on corn yield of, 72
  units of, 57-58
Soil organic matter, 223
  content in soils, 224
  determination of, 225
  relationship with N content, 224
  removal in texture analysis, 82
  removal in XRD, 308
Soil pH,
  buffer, 108-110
  critical, 101
  definition, 96-97
  determination of, 99-102, 103-106
  effect of air drying on, 19
  hydrolytic, 107
Soil porosity, 86
  determination of, 94
Soil profile, 2-3
Soil reaction,
  definition of, 96
  effect on crops, 98-99
  types of, 98
Soil separates,
  definition, 73
  types of, 73-74
Soil test,
  monitoring in, 44-45
  objectives of, 43-44
  preventive, 44
  samples in, 44
Soil texture,
  calculations in analysis of, 81
  definition of, 73-74
  effect on soils and plants, 74
  heavy, 74
  laboratory analysis of, 74, 79-81, 81-84
  light, 74
Specific gravity,
  apparent, 88
  relation to soil density, 87-88
  specific, 88
  true, 88
Spectrophotometry,
  absorption, 243
  basic laws in, 244
  definition of, 242
  double beam, 249-250
  emission, 243
  single beam, 249
Spectroscopy, 242

Sphalerite, ZnS, 219
Standard addition method, 39-40
Standard reagent, 53
Statistical analysis, 38
Sticky point, 107
Stokes' law, 78
Strishovite, 199
Submicroscopy, 49
Suction method in water determination, 65-66
Sulfanilamide, 152
Sulphur,
  adsorbed, 185
  available, 184
  content in soils, 184
  determination of, 184, 186-187
  effect of airdrying on content of, 19
  extraction of free, 188
  extraction of inorganic, 188
  free (soluble), 185
  inorganic, 185
  organic, 185
  source of, 183
Sulfuric acid
  in determination of C_{org}, 227, 229, 230
  in Kjeldahl, 143
Surface charge density, 115
Surface tension, 56
Summation method, 35

Target in XRD tube, 303-305
Tectosilicates, 199
Tensimeter, 69,70
Tensiometer, 69-70
Tension plate, 62
Thermogram, 331
Thermogravimetry, 329
Tissue testing, 45, 47
Total acidity, 98
  determination of, 238-239
  of humic matter, 240
  values, 239
Theoretical endpoint of titration, 53
Tourmaline, 212
Transmission coefficient, 245
Transmittance, 245
  relationship with concentration, 248
Tridymite, 199
True value, 2
  definition of, 27
  scientific, 27-28
Tungsten filament lamp, 250
Turbidimetry, 244

Ulexite, 212

Vermiculite,
  DTA identification of, 315, 319
  XRD identification of, 351, 352
Volumetric analysis, 53
  types of, 53

Walkley-Black method, 230-231
Water,
  available, 58, 62
  determination of, 56,
  dry mass % of, 59
  total, 58, 59-60
  volume % of, 59
  wet mass % of, 59
White bud of corn, 220
Wilting point, 58
  determination of, 62-65, 65-66
  pF of, 58
Wulfenite, $PbMoS_4$, 216

Xenon lamp, 251
X-ray,
  application of, 302
  coherent scattering of, 301
  diffractogram, 307
  energy of, 299, 304-305
  energy dispersive analysis by, EDAX, 303
  fluorescence, 301
  incoherent scattering of, 301
  secondary, 302
  tube, 304
  types of, 299, 300
  wavelength dispersive analysis by, 302
X-ray diffraction, XRD,
  first-order, 307
  pretreatments in, 309
  random oriented sample in, 312
  random powder technique in, 312-313
  second-order, 307
  third-order, 307

Zeeman effect, 261
  effect on element concentration, 262-263
Zinc, 219
  content in soils, 219
  deficiency of, 220
  determination of, 220-221
  effect on plants, 220
  exchangeable, 219
  extraction of
    available, 222
    exchangeable, 221
  oxidation of, 219
  source of, 219
  toxicity, 220